AF605765

FISHING THE TONGARIRO

A History of Our Greatest Trout River

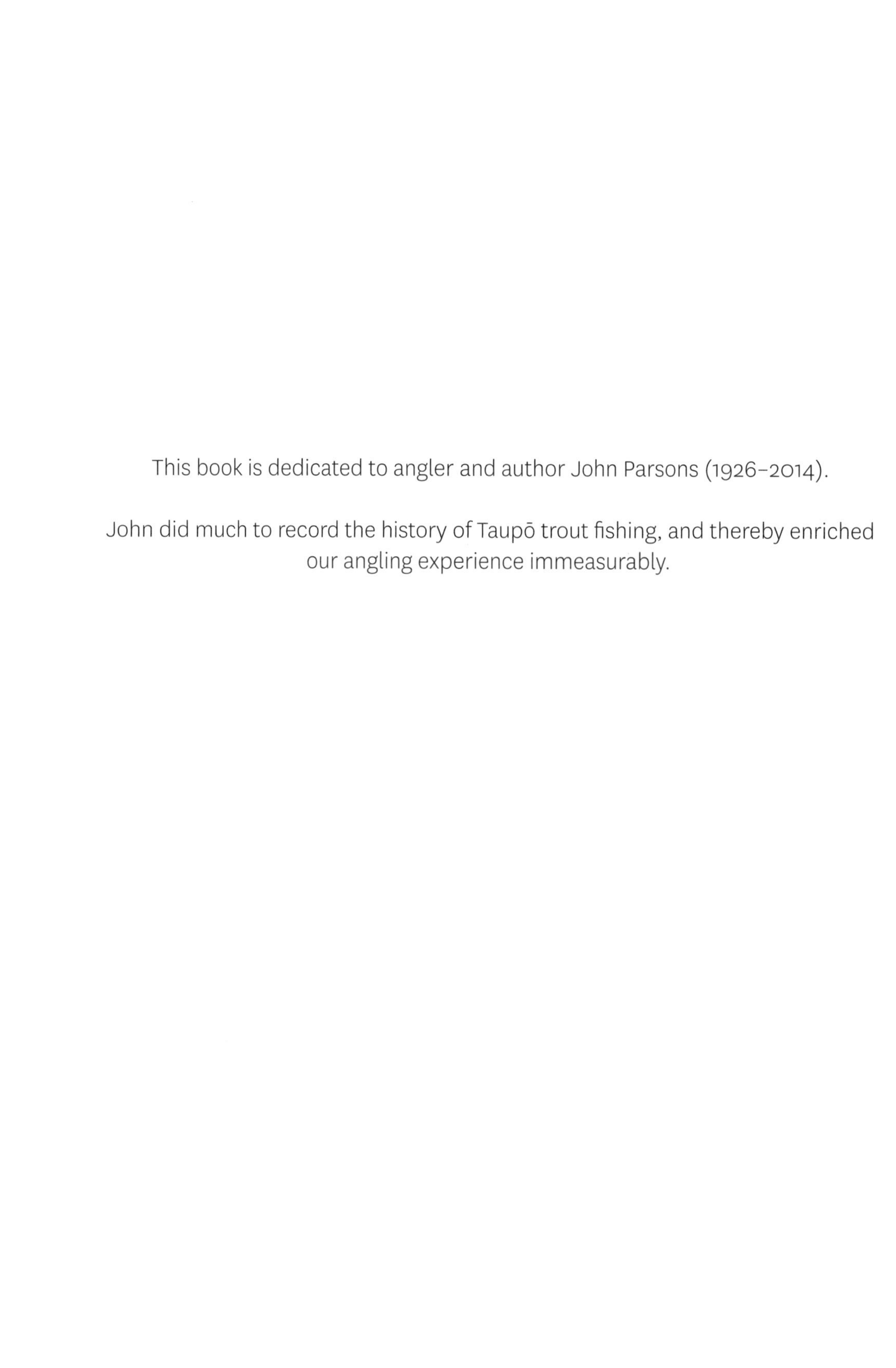

This book is dedicated to angler and author John Parsons (1926–2014).

John did much to record the history of Taupō trout fishing, and thereby enriched our angling experience immeasurably.

FISHING THE TONGARIRO

A History of Our Greatest Trout River

GRANT HENDERSON

Contents

Joe Frost with a fresh trout catch on the banks of the Tongariro.

Whites Aviation Ltd: Photographs. Ref: WA-10045-F.
Alexander Turnbull Library, Wellington, New Zealand.
/records/22354027

Preface

IN 1996, WITH DECADES OF EXPERIENCE of trout fishing in New Zealand's rivers and lakes, I contacted John Parsons to suggest he write a book detailing the history of fishing on the Tongariro River, complete with photographs. As a long-time Taupō resident, an angler and a local historian, John was ideally qualified for the task. He was good enough to approach a publisher with this concept, but was turned down because the potential market was thought to be only regional and therefore too small. After giving the idea further thought, he suggested that I should write the book, and kindly offered assistance in the form of photographs and files he held.

The book you now hold is the result. It is not of the 'where to go and where to get them' type of fishing book, nor is it a memoir. Instead, it puts the Tongariro River at its centre and weaves various historical strands around it in a kind of angling regional history.

It will be my eternal regret that John Parsons died before the book was published. However, we exchanged material and ideas over the years, I visited him on two occasions, and some of his photographs appear in the book. He was a keen student of the history of the Taupō fishery, and his kindness and enthusiasm for the project were among the things that kept me going when the task of writing seemed insurmountable.

One of the themes I discovered when researching the book was conflict. There has been conflict between people in the Taupō region ever since Europeans first arrived, conflict between fish species as the introduced trout took over and decimated the native kōaro, conflict between brown and rainbow trout as the latter became the favoured sports fish, heated discussion as to the proper name of the river, arguments between citizens and the government over the Tongariro power scheme, and even disputes between anglers over fishing techniques and angling etiquette.

Much has happened at Taupō since trout were introduced to the region some 120 years ago. Anglers have come and gone, new angling methods have been introduced, more fishing water has been opened up, and debates among anglers continue. But as John Parsons wrote in the very first chapter in his book *Parsons' Glory*, it doesn't matter if you don't catch anything. Hope springs eternal, and there will always be the magical thrill of simply going fishing.

Grant Henderson
Auckland, May 2023

CHAPTER 1:

The geography of the Tongariro River

We learn geology the morning after the earthquake . . .
— **RALPH WALDO EMERSON**

BEFORE DISCUSSING THE HISTORY of New Zealand's premier trout river, the first step must be to define the waterway. Where is its source? What course does it follow? Where does it end its journey? In the case of the Tongariro River these apparently simple questions have provoked controversy. The anglers' paradise known as the Tongariro River can be traced to the slopes of Mt Ruapehu on the North Island's central plateau. Here the Upper Waikato Stream begins. At various points on its journey this stream is joined by several other waters, one being the Waipākihi Stream which has its source in the Kaimanawa Range to the south-east. As the bodies of water grow larger, what were once streams transform to rivers.

The rivers in the Waikato catchment did not all emerge at the same time, nor did they always follow their current paths. Professor Paul Williams, a geoscientist at Auckland University's School of Environment, explains how the Waipākihi came to be a source of the Waikato:

> . . ., by far the largest volume headwater of the Tongariro is the Waipākihi River that flows from the Kaimanawas. The Upper Waikato is a much smaller and much younger tributary. The Kaimanawas are millions of years older than Ruapehu, so drainage from them was well established before Ruapehu volcano was initiated. However, the

> Waipākihi originally drained south and was diverted north to Taupō (hence the hairpin bend) when volcanic debris from the expanding cone of Ruapehu blocked its southern path. That diversion probably occurred around 100 to 200 thousand years ago (Ruapehu started to grow around 300,000 years ago).[1]

To the west, the Wharepu, Mangatoetoenui, Waihōhonu and Oturere streams rise on the eastern slopes of Mt Ruapehu, while further north there is a major tributary in the form of the Poutu River, the outlet from Lake Rotoaira.

One river, two names

Thanks to the arcane rules of mapmakers, the Tongariro River as we know it today is a segment of a greater river known as the Waikato. For official reference purposes the Waikato becomes the Tongariro at the point where it is joined by the Waihōhonu Stream, which flows off the slopes of Mt Ruapehu. But arguably the Upper Waikato Stream and Waihōhonu don't join together. Both can be seen as left-bank tributaries of the Waipākihi, which had its name changed at the hairpin bend referred to.[2] From there it flows north for some 55 kilometres until it enters the southern end of Lake Taupō, near the town of Tūrangi.

Rather confusingly, the name Waikato is also applied to the river which begins at the outlet on the northern side of Lake Taupō. One of New Zealand's longest rivers, its course runs for 425 kilometres before entering the sea at Port Waikato. The reference to Tongariro for the initial part of the waterway reflects the years of that name's use by anglers who fished the river. The official body which ruled on the issue, the New Zealand Geographic Board, specifically took their views into account when it made its decision.

Explorers surveying territory in the early days of European settlement in New Zealand sometimes adopted place names commonly used by local Māori, who were the first inhabitants. However, they did not always do so; often the name of an English town (Cambridge) or an aristocrat, politician or military hero (Nelson, Palmerston, Marlborough) or the sponsor of a colonial association would be transplanted to the Antipodes. But for rivers the indigenous terms seem to have been retained in most cases, examples in the South

Looking north on the Tongariro River above the Turangi road bridge, 1912, with Lake Taupo in the background.

Ref: PAColl-8645. Alexander Turnbull Library, Wellington, New Zealand. /records/22479630

Island being the Rangitata, Waimakariri and Waitaki, while in the North Island there are the Manawatū, Rangitikei and Whanganui.

In 1859 the Austrian Ferdinand von Hochstetter made the first geological survey of New Zealand under instructions from the New Zealand government. His map of the upper half of the North Island shows a river called the Waikato as far upstream as the hairpin bend (he didn't comment on where it rose) and heading north to join Lake Taupō's southern shore at Tokaanu. The name Tongariro is applied to another river rising in the Kaimanawa mountains to the east and which joins the Waikato.[3] Von Hochstetter's conversations with the Māori inhabitants revealed that they gave the name of Tongariro to the part of the river from the delta to its junction with a major tributary rising in the south-east[4], with the Waikato River losing its name for part of its course.[5]

The question of the correct name of the river was to come up for lengthy debate some 70 years later.

Tongariro geology

The Taupō region has a violent geological history. The lake was formed some 25,400 years ago in the Ōruanui eruption, which showered an estimated 800 cubic kilometres of pumice rock, ash and other material over a wide area.[6] This region is known as the Taupō volcanic zone.

The catchment area of the Upper Waikato River covers some 771 square kilometres. The Kaimanawa Range in the east is largely made up of greywacke, a hard, non-porous basement rock. Native forest grows here in thin soil cover, and includes trees such as beech. This area is steep and, despite the vegetation cover, there is a high degree of rainfall runoff.

To the west, near the central plateau mountains, lies a plain covered in tussock and mānuka vegetation. The geology of this region comprises young volcanic rock with deposits of pumice and breccia, the result of volcanic eruptions.[7]

Lahars a threat

Volcanoes in the central plateau have also influenced the landscape by expelling lahars from time to time. A lahar is a mixture of rock, sand, silt and water — rather like wet cement — which can flow from a volcano at very high speed. It may be caused by an eruption of lava, ash and hot gas (pyroclastic flow), the collapse of a crater lake or even heavy rain. Lahars tend to follow low ground such as stream channels, although rivers and streams do not always follow the paths carved out by lahars.

Geologists' studies indicate that for thousands of years lahars have emerged on the northern slopes of Mt Ruapehu. One was expelled down the bed of the Mangatoetoenui Stream, travelling as far as Lake Taupō's southern shore.[8] The Upper Waikato and Tongariro rivers have followed the lahars' course. A lahar also affects the Waipākahi River at the point where it emerges from a fault on the Kaimanawa Range. This river at first takes a south-westerly course through a deep valley, but turns sharply north-east thanks to ring-plain deposits from Mt Ruapehu and Mt Tongariro which have formed against the Kaimanawa Range foothills. These deposits, in combination with the lahar's softer sediment, blocked the Waipākihi River and forced it to make a u-turn northwards to join the Upper Waikato via the so-called hairpin bend.[9]

OPPOSITE: Map of Lake Taupō and the Upper Waikato (Tongariro) River, illustrated by Johannes C. Andersen.
James D Richardson, Auckland Libraries Heritage Images Collection

Te Aroha
Morrinsville
Hamilton
Cambridge
Oxford
Kihikihi
Lichfield
Tauranga
Te Puke
Maketu
Oropi
BAY OF PLENTY
Matata
Whakatane
Rotorua
Galatea
Huka Falls
Taupo
LAKE TAUPO
KAINGAROA PLAINS
RANGITIKEI RIVER
Reference
Boundary of Volcanic Zone
Hot Springs & Geysers
Roads
Railways
Railways under construction
PLAN No. 1
TAUPO
VOLCANIC ZONE
Scale of Miles

The Waikato Falls at Begg's Pool on the Upper Waikato River block the upstream passage of trout.
D. Nicholson, Archives New Zealand, Archway Item ID:R24748882, Archway Series Number: 6539

For the angler, the most obvious clue to the Taupō region's volcanic history is the distinctive nature of the riverbed — a mixture of large round boulders, coarse sand and silt.

In its upper reaches the Tongariro River flows through high-walled, narrow gorges before widening into a series of pools with fast-flowing rapids and runs. Once it reaches Tūrangi, near Lake Taupō, it broadens and slows. Sediment (rocks, gravel, ash and silt) carried down from mountains in the central volcanic plateau ceases to be carried by the river's flow and settles on the riverbed. The pools below the main highway bridge form wide, curving expanses of quieter water on the way to the delta, where thousands of cubic metres of sediment have been deposited for millennia.

Mt Ruapehu has disgorged lahars in 1895, 1945, 1969, 1974, 1975 and 1995–96, often with disastrous results for trout fishers. Department of Conservation scientists conclude that a lahar from Mt Ruapehu could potentially alter the quality of Tongariro's water

The Tongariro River divides into several mouths when it enters Lake Taupō at the delta.

Tongariro River, Lake Taupō, Waikato. Ref: WA-41413-F. Alexander Turnbull Library, Wellington, New Zealand. /records/30633995

by increasing suspended sediment concentrations, changing the acidity and chemical composition, and possibly raising the water temperature.

The timing and location of the event are also key factors. Assuming the Tongariro River catchment is affected to some degree, the greatest damage would occur in late spring and summer (November to March) when the maximum numbers of juvenile trout are present. Mortality would be less for a lahar appearing in April or June, when fewer trout of all age groups

The bluff at the Duchess Pool shows the Taupō region's volcanic history in layers of rock and pumice soil.

Ref: PAColl-8645. Alexander Turnbull Library, Wellington, New Zealand. /records/22479630

inhabit the river. On a brighter note, the scientists consider that, because lahars are of short duration, the Taupō fishery, aided by the Tongariro hatchery, can rebuild itself. Conditions suitable to rear young trout would arise after a year.[10]

The mountain's occasional eruptions are potent reminders of what is going on beneath the ground in the Taupō volcanic zone. In 1996 one such event produced an estimated 7 million tonnes of ash, with 2.3 million tonnes falling on Lake Taupō. Some ash entered the Tongariro River, with disastrous effects on the power generation equipment.

The relatively short span of human life tempts us to regard the Earth's crust as static, but in reality powerful subterranean forces are constantly at work. It is a distinct possibility that significant volcanic activity can occur, as shown by the crater eruptions from Mt Tongariro in August and November 2012. These events, which were entirely unexpected, caused no fatalities,

while unfortunately the same could not be said of the eruption at Whakaari/White Island in the Bay of Plenty in December 2019. Such disasters are clear evidence that while fly fishers at the Tongariro River may not be dancing on a volcano, they are definitely fishing near one.

Broad, winding channels characterise the Tongariro River as it approaches Lake Taupō. In times of flood it frequently breaks its banks and submerges surrounding farmland.
Alexander Turnbull Library, Wellington, New Zealand. /records/30631754

CHAPTER 2:

Establishment of the fishery

It is suggested that Lake Taupo should be stocked with trout. The idea would be a good one for the Auckland Acclimatisation Society to take up.

— ***THE WANGANUI HERALD*, 30 JANUARY 1893**

THE STORY OF HOW TROUT became established in Lake Taupō and its tributary rivers, including the Tongariro, is one of a few people who showed considerable vision and remarkable persistence. Lake Taupō is located on the central plateau of New Zealand's North Island, some 609 metres above sea level. In the latter half of the nineteenth century, European settlers seeking to work the land found it easier to establish themselves near coastal settlements in northern regions such as Auckland, with its busy port, meaning the Taupō region remained relatively sparsely populated.

However, this was no isolated backwater; the area abutted areas that were hotspots of action, and saw its fair share of activity, albeit less volatile in nature. In the early 1860s the Waikato region, to the north of Lake Taupō, was the scene of major conflict between the government and Māori, who refused to give up their sovereignty and accept a foreign monarch as their ruler. The Waikato War proved to be the major series of battles that broke the main Māori resistance and paved the way for European domination of the North Island.

Under their leader Horonuku Te Heu Heu Tukino, Ngāti Tūwharetoa had joined the Waikato tribes in their resistance to the government. Te Heu Heu also gave support to Te Kooti, who carried out a successful guerrilla war campaign in the central North Island, before being defeated in the shadow of Mt Ruapehu at Te Pōrere in 1869 and retreating west to the King Country. Te Kooti remained there for more than 10 years without engaging in further conflict, before receiving a pardon under a general amnesty in 1883.

With each successive battle in the Waikato region the Māori defensive line fell back southwards. The settler government followed up on its military successes with land confiscations to pay for the war — in late 1864 more than 1 million acres were taken in the Waikato, Bay of Plenty and Taranaki.[1] Once the conflict died down, the Native Lands Act 1865 put the final seal on the process by permitting communally owned Māori land to be converted to ownership by individual Māori, thereby enabling it to be sold to Europeans or, as was often the case, taken in satisfaction of debts the Māori owner had incurred for material goods.

By the 1880s the Taupō region was proving attractive to European investors eager to develop farms. Others thought there would be gold in the Kaimanawa mountains. In 1884 an English syndicate purchased 40,000 acres for £32,000, conducting the whole transaction by cable from offshore. As the land was allegedly subject to a Crown proclamation which forbade a direct sale to Europeans, the affair drew caustic comment from the press.[2] The Act had been intended to apply to a very limited amount of land containing exceptional thermal vents and other attractions (perhaps 4,000–5,000 acres in all), but it was instead applied to hundreds of thousands of acres in the central North Island where no thermal activity occurred, much of it in east Taupō.[3]

Whatever the route, land disappeared from Māori hands. By 1884, most of the Māori land on the eastern side of Lake Taupō had been alienated, but the areas to the west and south remained theirs, as did property bordering the Waikato River near Taupō.[4] This alienation was part of a wider trend, with Māori land ownership in the North Island halving from 22 million acres to 11 million acres in the years from 1861 to 1891.[5] By 1891, Māori comprised 10 per cent of the national population, but their land ownership had reduced to 17 per cent of the land.[6]

The role of the Armed Constabulary

The few Europeans who did arrive in the Taupō region in the 1850s were commonly a colourful mix of members of the Anglican Church spreading the Christian message, adventurous naturalists in search of new species for their collections and artists seeking to record the unexplored territory on canvas.

The earliest settler residents of the district were members of the Armed Constabulary, a force created to enable the government to penetrate the North Island's central plateau region and establish its authority in the region. By the 1870s, a number of these men had left the force and settled at the head of the lake in the village of Taupō, far from the coastal settlements.

The first of these European explorers sought to establish farms so they could live off the land. A sports fishery was far from their minds, even though the attractions of the area — mountains and thermal activity — were obvious enough as a potential source of tourist income. In 1862, J. C. Crawford, a Wellingtonian, had remarked on the potential of Lake Taupō and its surrounding countryside for fish and game. The lake was New Zealand's largest, at 616 square kilometres it was fed by many rivers, had no fish species larger than an index finger, and was not populated by eels; all factors which pointed to a perfect sports fish habitat.

Trout introduced to New Zealand

Given the extreme isolation of the Taupō district, it is not surprising that trout acclimatisation first occurred in other districts. In the South Island, several attempts by the Canterbury Acclimatisation Society in the 1860s saw brown trout ova brought from Tasmania to the city of Christchurch.

The ova had been obtained by James Youl, acting for the Tasmanian Salmon Commission. The first trout — 12 in all — were successfully reared in ponds on the Derwent River near Hobart on 4 May 1864. New Zealand's brown trout are direct descendants of these first few specimens. Because New Zealand acclimatisation societies had contributed to the costs of importing the ova from the United Kingdom, they were entitled to receive some of the ova produced later on for their own use. In August 1867, an official of the Canterbury Acclimatisation Society duly uplifted 800 ova from Tasmania. The shipment was to be divided equally between the Canterbury and Otago societies. However, the importation was hardly a success, with only three of the ova hatching.[7] Remarkably, a male and female trout from this small group became the source for all brown trout subsequently raised in Canterbury.[8] Further shipments of ova took place in 1869 and 1870, and trout raised from these imports were sent to many parts of New Zealand. By the 1880s, the species was flourishing, and reports of large catches of sizeable trout were a regular occurrence.

Residents in the thermal districts of Rotorua and Taupō saw no reason why the process could not be replicated in their districts, and so the Rotorua lakes were stocked with brown trout in the 1880s by the Auckland Acclimatisation Society. Of course, at this time the brown trout — 'the gentleman's trout in Harris tweed and plus-fours' according to some — was seen as the only trout species worthy of consideration.

From the earliest days of settlement, residents of the colony had been determined to make the New Zealand countryside a mirror image of England when it came to game animals. The early colonists therefore enthusiastically released all manner of mammals, birds and fish imported from the Mother Country on a trial-and-error basis, from the 1860s often under the auspices of the acclimatisation societies. Species let loose in competition with New Zealand's indigenous animals and fish included red deer, sparrows, pigeons, pheasants, quail, waterfowl, salmon and trout — anything thought worthy of the sportsman's gun or to provide food for the table. Other species considered for introduction into the country had less recreational appeal. In 1898 the Auckland Acclimatisation Society decided to apply funds towards the importation of bats and toads — British ones, no doubt.[9]

New Zealand's acclimatisation activities were carried out with one important social objective: to make fish and game available to all strata of New Zealand society. The division between the laird and his gamekeepers, based on private ownership of game animals, would not be transported from the Old Country to New Zealand. Instead, fish and game would be accessible by anyone who paid a licence fee and was skilled with a rod or a gun.

Taupō's native species

The first Europeans to reach Lake Taupō found its waters harboured three species which provided a source of food to Māori. These were the freshwater crayfish, or kōura, the freshwater mussel, or kākahi, and a small fish called kōaro (*Galaxias brevipinnis*). The native New Zealand grayling was not resident in Taupō waters, but kōaro were an important food for Māori, who had a detailed knowledge of their migratory habits and knew the best time to fish for them.

Kōaro have an intriguing lifecycle. The species is able to colonise inland lakes at the juvenile stage, thanks to its specially formed pectoral fins that have adapted to climbing steep rock faces and walls. Because Lake Taupō kōaro are landlocked, they use the lake as their ocean, spawning in tributary streams. They enter the lake at the juvenile stage and form shoals there, at which time — post-European settlement — they would have been exposed to heavy trout predation. They were also common in Lake Rotoaira, to the south-west of Lake Taupō.[10] They would spend part of their time in the underground waters feeding that lake, before emerging in October to migrate via the Poutu River. It was at that time that Māori set traps for them, then either cooking their catch immediately or drying them in the sun for consumption later.

Although the introduced brown and rainbow trout reduced the kōaro population significantly, the latter were not completely eliminated, even though the Māori fisheries based on them collapsed. Researchers have concluded the kōaro populations stabilised at levels in line with trout predation.[11] When smelt were added to Lake Taupō in the 1930s, that species became the trout's main food item, in effect saving kōaro from concentrated trout predation. Kōaro can still be seen in Lake Taupō, with adults reaching 180 millimetres in length.[12] However, in some lakes smelt not only compete for the same food and space as kōaro, but also prey on them.

Introduced fish

The first fish species the Europeans liberated in the lake were not trout but coarse fish, in the form of golden carp and the North American whitefish. Whitefish are a member of the salmon family that do not require running water in which to spawn. They are bottom-feeders, and easily caught on rod and line. When the New Zealand government secured whitefish ova from the United States in 1880, J. C. Firth of the Auckland Acclimatisation Society took 250,000 ova to Taupō — a long and arduous journey — and placed them 'in suitable places in the lake'.[13] A little more than a year later, the society noted that no information had been received on the results of the experiment, and concluded that the whitefish had failed to acclimatise in Taupō's waters.[14] Had they done so, the biology of Lake Taupō in the twentieth century would have been very different, because the kōaro might well have remained in large numbers.

The carp had a limited success rate. The numbers stocked in Lake Taupō were as few as 12 on one occasion in 1872,[15] and they were not often seen, although the lake was said to be well stocked with them by 1890.[16] Māori named the fish 'morihana', after the man instrumental in liberating them, Captain Morrison.[17] Carp may still be seen in some parts of the fishery today, Lake Rotongaio being one location.

In early July 1887 a Taupō resident, one J. Ross, asked the Auckland Acclimatisation Society for Lake Taupō to be stocked with either trout or catfish. The society responded that it had no trout for distribution at that time, but generously offered to send catfish at once, with trout to follow later.[18]

Attempts to introduce brown trout into Lake Taupō began thanks to Taupō resident Major David Scannell, a career soldier who came to New Zealand in the 1870s after helping to put down the Indian Mutiny. He joined the Armed Constabulary based at Taupō, and settled there when his 12 years of military service were complete.

Scannell built a fish hatchery at Nukuhau, close to the place where the Waikato River leaves the lake on its long journey northwards to the sea. In 1884 he informed the Auckland Acclimatisation Society that he was willing to rear trout ova if they sent him some. At first the society felt this suggestion was impractical because Taupō was too far away. However, the society must have changed its mind, for in 1885 it sent Scannell 1,000 brown trout ova[19] which he tried to rear in the cool, spring-fed waters of a small tributary; an ideal environment for juvenile trout. Nevertheless, it seems that this attempt ended in failure.[20]

Another contributor was Daniel McDonald, who reared trout at his farm, Lochinver Station, some 30 miles to the east of Taupō town using ova sent by the Hawke's Bay Acclimatisation Society. In 1887, McDonald liberated 24,000 trout fry in streams entering the eastern shores of Lake Taupō.[21] If the combined efforts of Scannell and McDonald achieved any measure of success, then it must have been so limited as to have passed unnoticed. In 1894 a correspondent informed *The New Zealand Herald* that, contrary to rumour, trout could survive in Taupō waters, and referred to 'an unsuccessful attempt . . . made to stock the district eight or nine years ago' — probably a reference to Scannell's efforts.[22] What is certain is that in the next few years Lake Taupō appeared to contain no trout worthy of an angler's attention.

The apparent failure of the 1887 liberations, assuming the fish were properly managed before their release, is something of a mystery. In terms of food supply, water temperature, absence of competing species and availability of nursery streams, the Taupō region favoured trout propagation in every way. Even more striking is the way in which trout had earlier been successfully established in Canterbury rivers, where predators in the form of eels were common.

Gavin Park's efforts

The next liberation of brown trout in the Taupō region was again courtesy of the Hawke's Bay Acclimatisation Society. This body provided ova to Gavin Park, the Taupō postmaster, who duly reared some 12,000 fry.[23] On 16 December 1892, C. A. Fitzroy, secretary of the society, liberated 6,000 brown trout in a Tongariro tributary. The Puketarata Stream is some 14.5 kilometres south of Tokaanu and a short distance from the confluence of the Tongariro River and the Poutu Stream. Fitzroy wanted to mark this event for posterity, for on the same day he noted the release in the visitors' book of Blake's Hotel at Tokaanu, no doubt with a touch of pride.[24] (Another source refers to the liberation being made in the Tongariro River, but Fitzroy's contemporary account carries more weight.) The pioneers also had their eyes on Lake Taupō's wide waters, because they kept 200 fry to be used as breeding stock for releasing into the rivers feeding the lake.[25]

This liberation must be accorded some degree of success, because Tokaanu residents reported sightings of trout in the main river four years later. The year 1893 is generally accepted as the date of the first Lake Taupō trout liberations, the source being 20,000 eyed ova sent by the Wellington Acclimatisation Society.[26] The following year Gavin Park and some local volunteers raised 80,000 brown trout fry from 110,000 ova sent by the Wellington Acclimatisation Society. Their efforts led to fry liberations in the rivers bordering the eastern side of Lake Taupō, such as the Hinemaiaia, Tauranga-Taupō and Waitahanui.

Samuel Crowther, a local man operating the coach service on the Taupō–Napier highway, made these liberations. Two others, John Grace and William McCauley, are recorded as making liberations in the Tongariro River — the first to be made in that waterway.

In 1895 the New Zealand government made a grant of £100 to the Wellington and Hawke's Bay acclimatisation societies for use in sending trout ova to Taupō.[27] The Wellington society arranged a shipment of 100,000 ova as a result.[28] Despite an earthquake at Taupō soon after the ova arrived, Postmaster Park's efforts saw most of the eggs hatch. By January 1896, trout were such an obvious presence that they caused comment in local newspapers: 'Trout are plentiful in the Lake and in all streams round about,'[29] said one; 'Every stream around the lake is now thoroughly stocked, also the Waikato river,' announced another.[30]

Although Gavin Park's voluntary efforts for the angling community passed largely unnoticed, he was to receive publicity — albeit of the wrong kind — several years later. A married man with seven children, he had been with the Post Office for 30 years, but somehow fell into debt in middle age. When pressed by creditors, he stole £442 which came into his hands as postmaster. Although he intended to repay this sum, a surprise audit uncovered the cash deficit before he had done so.

Park tried to divert the investigation by suggesting he and the inspector take a walk, but when that failed the theft was sure to be uncovered so he promptly confessed. In March 1896, he pleaded guilty to a charge of theft from the Post Office Savings Bank. The conviction meant he forfeited his £520 retirement pension, so the government actually made a gain. The court concluded that the loss of both his pension and his reputation were sufficient penalty, and sentenced him to 12 months' probation instead of a gaol term.[31] That was seen as a light penalty, with one newspaper describing it as lenient in the extreme,[32] but — who knows? — perhaps the judge was a fisherman.

Brown trout established

By 1898, the cumulative stocking process had had the desired effect, with brown trout now a common sight in the Tongariro River near Tokaanu. More importantly, they were being fished. Trout fishing thus became a new tourist attraction in the district. Visitors could try their hands at fishing, view the local mountain scenery, bathe in a thermal pool at Tokaanu, and then head south to take in the beauty of the Whanganui River — a scenic waterway navigable by steamer and touted as 'the Rhine of New Zealand'.

The identity of the first person to fish for the brown trout immigrants will never be known for certain. One candidate is Ernest de Lautour, a diehard angler who, living in spartan conditions at the riverside, had the opportunity to study the browns' habits. He observed that while the brown trout thrived on a diet of fish, they were too well fed, and generally took the lure only at night. In 1900, a correspondent wrote:

> We are still tantalised by seeing shoals of fine large trout in the lake daily and unable to catch them. They will not take any kind of bait which shows that they have ample feed in the waters of Taupo Moana. Some of the fish are very large. About twenty five in one shoal would nearly all have turned 14lb to 16lb.[33]

This uncooperative species required the use of live bait to ensure success; a matter referred to by tourists Malcolm and Forrestina Ross, who visited de Lautour at Tokaanu in 1898. Captain Thomas 'Darby' Ryan also commented in later years on the difficulty of taking the browns on the fly.[34] With an air of desperation, a newspaper reported that the trout in Lake Taupō would not take the rod or line, and a proposal had been made to allow netting.[35] Even so, from this time forth trout fishing was a major drawcard for visiting tourists, with brown trout being caught at impressive weights of 12–18lb. The current average weight of Tongariro River trout is 4–5lb, although the occasional monster can be taken. In March 2017, for example, Aidan Wildey caught a brown trout weighing 15lb 4oz from the river's middle reaches.[36]

A little to the north of Taupō, in the thermal lakes district, a significant advance then occurred. Lake Rotorua anglers were similarly convinced that trout in their lake would not take a fly. Charles Morison, a Wellington lawyer,[37] is recorded as the first angler to take Rotorua trout on fly tackle, during Christmas 1903, thereby disproving the pessimistic views of locals. W. H. Tisdall, the Wellington fishing tackle dealer, then followed up with an impressive bag of rainbows, the largest 12lb, taken in a gale near the mouth of the Ngongotahā Stream.[38]

Gradually, anglers at Taupō began to score some successes with the reluctant browns. Very early in the twentieth century, it soon

Tokaanu, Lake Taupō, from Munganamo, King Country.

Te Papa Registration Number C.016311

TROUT FISHING SEASON IN NEW ZEALAND: A FINE BROWN TROUT CAUGHT IN THE WAIKATO RIVER AT THE ENTRANCE TO LAKE TAUPO, AUCKLAND.

This fine specimen, which was presented by Mr. Gallaher, of Taupo, to the Government Tourist Department, weighed [illegible]. Its length was [illegible] inches, and girth [illegible] inches.

Brown trout, weight 21½lb, 33½ inches long.

Auckland Libraries Heritage Collections AWNS-19051221-17-3

became clear that for those willing to endure the difficulties of night fishing, there were exciting rewards to be had. Mr R. Chase, a Tokaanu resident, wrote in the press in 1906:

> I started fishing here in the beginning of the season, and up to date have caught 311 brown trout with an average weight of 5lb. I only fish at night. Plenty of people passing through here go out for an hour or two in the daytime, and, because they fail to land a fish, go away and tell their friends there is no sport here.[39]

Others joined him. From early January 1906 to late April 1906, a resident Tokaanu angler caught 76 trout in the Tongariro, at a total weight of 925lb, averaging 12lb each. His best catch in one day comprised four trout weighing 12lb, 16lb, 19lb and 25½lb.[40] At the same time, the rainbows were starting to appear in anglers' catches, with reports of fish around the 10lb mark being taken from the river at Tokaanu. Nearly one year later, a brown trout weighing 29lb was taken at Tokaanu on a minnow.[41]

In 1909, another report noted:

> Tongariro river, Taupo, has again proved itself our best-producing river for big fish. Mr. Sam Crowther sent down this week a magnificent brown trout weighing 20lb, [which] was on view at Tisdall, Ltd.'s, sports emporium on Lambton-quay. Mr. Crowther's letter accompanying the fish states that he hopes before the season closes to send down something even better than a 20-pounder. Undoubtedly Taupo and the vicinity produces the largest fish in New Zealand.[42]

The arrival of the rainbows

The acclimatisation societies of the central North Island ultimately became dissatisfied with the sporting limitations of the brown trout. The Auckland body found the species had not established itself in local waters — possibly because the water was not cold enough — and turned its attentions to building a hatchery near Lake Taupō.[43] If the North Island thermal lakes district was to fulfil its potential as a sport fishing centre, another trout species was needed; one which came more willingly to the fly and which could be taken in daylight. In casting around for a more desirable substitute, the societies settled on *Salmo gairdneri*, the North American rainbow trout, now called *Oncorhynchus mykiss*.[44] Attempts at importing the species began in the late 1870s, but it was not until 1883 that 30,000 ova, thought to be steelhead (sea-run rainbows), were sourced from Sonoma Creek in California.[45] Steelhead are not a different species; they are simply rainbow trout which are anadromous, that is, they migrate to the ocean to attain their adulthood, after spending their first years of life in the river where they were born.

It seems that rainbows arrived in New Zealand accidentally, because the acclimatisation society had asked for a shipment of

Sir Cecil Whitney, the man whose foresight and drive brought rainbow trout to Taupō waters.
Grey Whitney

brook trout. Some 5,000 rainbows were among the shipment that was eventually hatched, although their presence as a separate species was not immediately obvious. The rainbows reared by the Auckland society are the first of their kind successfully bred in New Zealand,[46] a matter now formally recorded with a plaque at the pond in the Auckland Domain gardens.

Auckland businessman Cecil Whitney later stated that the rainbow ova were a gift to the Auckland Acclimatisation Society from a United States visitor, Mr Daniels. The gift was a mark of thanks for the hospitality Daniels received (apparently in the form of pheasant shooting).[47]

An unofficial trout liberation took place at Lake Rotorua, some 80 kilometres to the north of Taupō, in the heart of the tourist thermal district. In 1893, Cecil Whitney and Captain A. Bull put rainbow fry into Fairy Springs and other streams entering the lake. The liberation was an unprecedented success, for by 1896 rainbow and brown trout were reported as thriving in some Rotorua lakes. However, the prevalence of poaching and the lack of control by the Tauranga Acclimatisation Society led the Auckland Acclimatisation Society to apply to the government for permission to take over administration of the Rotorua district fisheries. Rotorua residents decided to join the Auckland body in March 1895.

When the government granted the society's request and gave it jurisdiction over Rotorua County, the district of West Taupō was also included. However, the East Taupō county district was absorbed by the Hawke's Bay society, in effect giving it control over most streams entering Lake Taupō and at Tokaanu.[48]

Some six months later, in late February 1898, the Wellington Acclimatisation Society shipped 5,000 rainbow trout fry to the Taupō region. The fry came from the hatchery at Masterton, and travelled in the care of Frank Lowe, assistant secretary of the society, and two journalists, the husband-and-wife team Malcolm and Forrestina Ross. While the group's main objective was to climb the mountains of the central North Island plateau, they found time to boost the cause of trout acclimatisation as well.

A train trip northwards from Wellington brought them to the small settlement of Waiouru, where they made their first liberation in a dam near the stables. The party then headed for Lake Taupō in a wagon, with Forrestina Ross being given the honour of liberating the first fry in Taupō waters at a ford on the Upper Waikato Stream.[49] This location was above a nearby waterfall — the point where the Desert Road crosses the headwaters of the Upper Waikato.[50] Other tributaries which received the young trout included the Te Piripiri and Mangatoetoenui streams[51] which run off the northern slopes of Mt Ruapehu to form the Upper Waikato River. The trip northwards ended at Tokaanu, where the Rosses fished for the elusive brown trout in the Tongariro River. They had no takes, which simply underlined the need for a more catchable trout species in Taupō waters.[52]

That left Lake Taupō's wide waters seemingly devoid of rainbow trout, although the Auckland Acclimatisation Society made small liberations there in 1899.[53] The Hawke's Bay Acclimatisation Society received rainbow fry from its Auckland counterpart in 1901 and 1902, but made no liberations in Lake Taupō,[54] even though the brown trout there were proving hard to take on rod and line.[55]

Cecil Whitney intervenes

Enter Cecil Whitney. Few people can have done more for the acclimatisation of game animals and fish in New Zealand than this forceful, energetic Scot. Whitney joined his father's Auckland munitions firm, the Colonial Ammunition Company, in 1885, and worked his way up to become chairman of directors. During his 61-year involvement with acclimatisation he was at times seen as autocratic and overbearing, but he certainly got results.

Perhaps Whitney had heard that rainbows had successfully been established in the Whirinaki River at Galatea, to the south-east of Rotorua, in 1897.[56] If they could thrive in that locality, why not at Taupō?

By the end of the nineteenth century, the residents of Taupō could see the potential benefits from promoting the region as a trout fishery. Captain Ryan sought rainbow trout fry from the Auckland Acclimatisation Society,[57] which was at first reluctant to act, given Taupō's remote location (the region was outside the area under the Auckland body's control) and the comparative ease of stocking the Lower Waikato River and other streams closer to Auckland. The Auckland society acted only after Cecil Whitney, then on its fish committee, learnt of the plans for stocking the lake. He arranged for an allocation of trout fry and ova, which the Rotorua Motor Coaching Company generously offered to collect at the Rotorua railway station and convey immediately to Taupō. So it was in the spring of 1900[58] that the first effective introduction of rainbow trout into streams feeding Lake Taupō occurred through the efforts of Captain Ryan and other Taupō residents, backed by Cecil Whitney and the Auckland Acclimatisation Society.[59]

First rainbow trout caught

According to official records, the first rainbow trout taken from Lake Taupō on rod and line was caught in April 1904. The trout weighed 2.98lb (1.36kg).[60] However, there is evidence that the species had appeared in anglers' catches before that date. In December 1902 a report noted that a few lucky anglers were catching large rainbow trout in the Tongariro River. A few months later, in May 1903, three more were taken from the Waikato River near Taupō, the largest weighing in at 7½lb.[61]

Another report refers to an earlier catch by the Englishman George Bertram Holdsworth. Holdsworth's family had been active in wool textile manufacturing in Yorkshire for generations, and as a young man he travelled abroad for the firm, first going to Europe and then to

Cecil Arthur Whitney, in his office at the Colonial Ammunition Company, Auckland, in 1902.
Auckland Museum Collections PH-NEG-14966

the Antipodes. When he came to New Zealand in 1902, he found time for trout fishing. Years later he wrote of catching his first rainbow trout, an 18lb fish, from Lake Taupō in 1902.[62] If that trout was introduced to the region as part of the 1898 liberations, it must have grown extremely quickly if it weighed 18lb four years later. However, Lake Taupō was virgin water brimming with a variety of food items for the hungry rainbows, making it conceivable that some fish could easily have attained double figures in four years. For example, a male rainbow trout caught in Lake Taupō five years after being released as a fingerling weighed 21lb.[63]

The only other explanation is that Holdsworth caught a brown trout and mistook it for a rainbow. However, it seems he was no angling novice and so unlikely to make an error of that kind.

The Māori response to trout

To Māori, the desire of Europeans to take fish for sport must have seemed odd. Eels and whitebait were food items found in the environment and could be taken with a spear or net, as could native birds such as the wood pidgeon (kererū). Catching fish using a long stick and a lure made of feathers was an entirely different approach. Nevertheless, some Māori like Ngamotu Wiremu of Taupō took up trout fishing, and used the same fly-fishing and spinning methods as other anglers, as prescribed by the regulations.

The depredations trout caused to native species raised the argument that Māori could fish without payment of a licence fee under the Treaty of Waitangi. This was a common theme in court prosecutions brought for illegal fishing in the thermal lakes district at Rotorua in the early twentieth century. The trout wiped out crustaceans like kōura, creatures on which Māori had originally relied for food. Trout fishing regulations meant Māori had to pay to fish in their own waters, whereas formerly they had not. To them it seemed that the introduction of trout had worked an injustice.[64]

The acclimatisation societies had to admit that Māori at Rotorua had lost out, even though trout had been introduced after the Treaty was signed. Nonetheless, the societies were against allowing all Māori to fish for trout without paying a fee because they feared the fishery would be wiped out. The courts generally accepted the view that the Treaty of Waitangi did not apply to introduced fish.

In 1908 the government gave some recognition to the Māori plight. An amendment to the Fisheries Act 1908 allowed 20 full-season licences to be issued authorising a person to fish for trout for consumption by themselves and members of their family, but for no other purpose whatsoever. The fee for this grudging concession was 5 shillings.

Te Arawa continued to press their claim to ownership of the beds of Rotorua district lakes. Litigation dragged on in the courts for more than a decade, until the government reached an amicable settlement with the tribe in 1922. The agreement saw more fishing licences granted to iwi members, greater protection for kōura and an annual payment of £6,000.[65]

More trout liberations

The enormous size of Lake Taupō and the multitude of rivers feeding

Plump, well-fed rainbow trout taken from Lake Taupō at the beginning of the 1909–10 season. By that time these fish were achieving record growth rates.

Auckland Libraries Heritage Collections AWNS-19091125-14-4 Creator: Smith, The Auckland Weekly News 25 November 1909

THE FISHING SEASON: FINE CATCH OF RAINBOW TROUT AT LAKE TAUPO.

The trout at Lake Taupo, North Island, N.Z., are considered, by anglers, as fine as any in the world, and the lake is becoming a very popular fishing resort now that the Main Trunk line makes it so easily accessible.

Smith (Wellington) Photo.

it must have made the region's inhabitants feel that the first liberations were insufficient, so in 1904 they petitioned the Auckland Acclimatisation Society for 1 million rainbow fry. The society made further allocations in 1904–1905, but not in the numbers the Taupō locals wanted.[66] The ova they did receive were hatched under the care of Captain Ryan, who liberated 220,000 fry in 1905 and 1906.[67] The Rotorua postmaster, Roger Dansey, also liberated American brook trout near Taupō.[68]

By 1905 the trout were advertising themselves in spectacular fashion, as a lady visitor to Tokaanu recorded:

> By the evening — the most perfect evening — we were at the river's brink waiting for the rise. The west was crimson and gold, the

> mountains deeply purple, and the moon, half veiled with rosy clouds, was reflected in the river. At last the surface began to dimple. More and more rose, until one longed to have many eyes to watch the sight. At our feet, in the middle of the river, by the opposite bank, the fish splashed, circles intermingling with one another, until one's head got tired of twisting and turning. Some enormous creatures rose, great streaks of silver, like a little geyser, and fell with a resounding smack . . . It was a mystically beautiful scene, and a most interesting experience — although the fish scorned all the inducements our fishermen could offer them.[69]

The introduction of the rainbow species provided the impetus the Taupō region needed to come into its own as a sport fishing destination. By February 1907 newspapers reported catches at Tokaanu which included rainbows weighing 10lb or more.[70] Massive brown trout of 20lb-plus were still being banked, but the rainbows were now a permanent presence. A most attractive feature of this new species was its willingness to take a fly or spoon in daylight. In November 1907, Captain Ryan told the Tourist Department that the Tongariro river was full of rainbow trout: 'anyone can fish there in day-time without bothering for the brown trout at night as in the past.'[71]

The sporting reputation of the massive rainbows spread quickly, luring anglers from the far side of the world. The first golden era in Taupō fishing had begun.

CHAPTER 3:

Catching trout by the ton

Brown and rainbow trout are being caught in the Tongariro river and Tokaanu stream in large quantities, weighing from 10 to 20 pounds, with fly and minnow. Fishing parties are arriving from all parts . . .

— ***MANAWATU EVENING STANDARD*, 20 DECEMBER 1907**

THANKS TO ACCLIMATISATION SOCIETY liberations, brown trout were thriving in Lake Taupō by early 1899. From time to time a large fish would be found dead on the lake edge, while others would cruise in shoals near the shoreline; well within casting range for anglers, but, as we have heard, unwilling to take the lures so keenly presented to them in daylight hours.

At this time a few people began to explore what sport Tokaanu could provide. Thus it was that in early 1903 Manawatū angler C. W. Grove made a successful trip to the Tongariro River, where he caught deep-bodied browns weighing 10–14lb.[1]

While small quantities of rainbow trout were first introduced into Lake Taupō in 1900,[2] the main liberations took place in 1903.[3] As was the case with Lake Rotorua, rainbows soon became the dominant and most visible species in the lake, growing to almost the same size as their brown cousins. Rainbows were much easier to catch, coming readily to fly, spoon or minnow. They also had more spectacular fighting qualities, frequently leaping when hooked, whereas large brown trout preferred to bore deep and engage in a long tug of war.

News of New Zealand's superb hunting and fishing soon got back to the Old Country, as the United Kingdom was then called. Sporting journals told of abundant deer, pheasant, quail and duck, while for the angler there was deep-sea fishing for sharks and swordfish and unrivalled trout fishing. In Britain, hunting and fishing of comparable quality was affordable only to the wealthy upper classes, but in New Zealand it was open to all. The cost, too, was moderate; an Englishman marvelling that a licence fee of £1 bought a man the freedom of an entire province for deer shooting, while another £1 gave free access to the very best trout fishing in scores of rivers:[4]

> . . . with a small outlay 'Squire Lackland' can enjoy better sport, have more freedom and variety, and have a far better time generally in New Zealand than if he spent thousands in rent and expenses, or owned the best deer forests and trout streams in bonnie Scotland.

The big fish

Not only were fishing licences pleasingly cheap, but the bounty to be had could also bring a gleam to the angler's eye. Fish size in a wild trout population is represented by a pyramid; with the largest specimens being at the apex, but fewest in number. But for the lucky, they were soon there to be had, with newspaper reports of very large fish beginning to appear with some frequency in first decade of the twentieth century. These triumphs come down to us almost solely in print, as very few anglers bothered to take photographs of their catches in the early 1900s, private camera ownership still being relatively rare. Tokaanu's remote location in the central North Island also meant it received few visits from newspaper photographers, so the opportunities to verify the anglers' claims were lost.

A positive plethora of giants

In December 1905 a brown trout weighing 21½lb (9.75kg) was caught in the Waikato River outlet at Taupō. The following year, the Tokaanu expert Robert Jones took a 25½lb (11.5kg) trout,[5] while in 1907 A. G. Campbell is credited with one of 23¼lb (10.6kg)[6]. Diagrams of the Tongariro River trout were displayed in a Whanganui shop also in 1907,[7] the larger fish having fallen to the expert casting of Robert Jones on a fly. In January 1908 four large Tongariro trout, one weighing 20lb

(9.07kg) were displayed in a shop window in downtown Wellington.[8]

The brown trout gave the best chance of a fish in the 20lb (9.07kg) class. Two newspaper reports of anglers' catches indicate the probable chances of taking a trout of that size. The first records a Tokaanu angler who fished consistently during January–April 1906 and caught 76 trout, averaging 12lb each. Two were said to be 20lb or more.[9] Fishing the Tongariro in January 1907, a Whanganui angler, Woodley A. Prouse, recorded fish from 13lb to 21½lb, and many lesser fish from 8lb to 10lb — 'a positive plethora of giants'.[10] The fish were all taken from the river at night, the only time when the browns would take a lure.

The flies used to catch these massive trout imitated the kōaro. Colonel Parry, a visiting English angler, noticed the locals fished at night with a large fly more than 5 inches long, a size that would have closely resembled the small bait fish.[11]

In 1909, F. A. Nesbitt from Napier caught a rainbow weighing in at 21lb. With a length of 34 inches and girth of 23 inches, it was thought to be the largest rainbow taken from Lake Taupō to that time, beating Colonel Moore's fish taken the day before by a mere 4oz.[12]

While few photographs of these monster trout exist, *The New Zealand Herald* featured four (allegedly) 20lb trout in March 1911,[13] while a 20lb fish is photographed among those taken by English anglers Charles E. Lucas and Gerald Horlick a year later.[14]

The experience of Horlick's party on the Tongariro indicates the superb quality of the fishery at that time. Fishing over a 17-day period, 4–20 March 1912, his party of four landed 197 trout at a total weight of 1,610¼lb. Horlick's bag of 26 had the lowest average weight at almost 7½lb, while Charles Lucas caught 70 fish at an average weight a little over 8½lb. The bulk of the trout (151) were taken by fly fishing, with the Durham Ranger and Jock Scott patterns accounting for most of them.[15] Only three of the fish caught were browns, which shows how emphatically rainbows had displaced the other species, at least in terms of catch rates.

The largest trout reported as taken from the Tongariro, and certainly Taupō's biggest, was a 30¼lb (13.72kg) rainbow said to have been banked by a Mr Kelly on 7 May 1909. It measured 42 inches long by 30 inches in girth.[16] There is no photograph, but the fact that measurements were taken gives credence to the claim. Earlier that year, a rainbow of 17½lb featured in a news report as falling to the rod of Webster, an English angler fishing at Tokaanu, although it is unclear whether it came from Lake Taupō or the Tongariro River.[17]

Trout taken from the Tongariro River with minnow.

Auckland Libraries Heritage Collections AWNS-19110518-9-2; *Otago Witness*, 6 Nov 1912 p 45

Trout of this size were truly exceptional — and most probably the Taupō trout had the heaviest average weight in the world at that time. And for several months each year, a small group of people had the fishery all to themselves in what must have seemed like the world's most remote angling outpost.

Getting to Tokaanu

In the early twentieth century, New Zealand's isolation made short-duration fishing trips out of the question for foreign anglers. A visitor from the United Kingdom had to spend a month or more at sea, followed by a two-day journey just to get to Lake Taupō. So the angler would really need a lengthy stay at the Tongariro River to make the trip worthwhile, with always the chance that fishing conditions might not be the best. New arrivals at the lakeside settlement of Tokaanu (a Māori name meaning 'cold stone') might time their visits too early, with the trout not having yet started their annual run upstream. The river might be in flood and completely unfishable, or drought conditions might mean low river flows and a delayed spawning run.

Fortunately, trout were not the only hard-fighting game fish on offer in the Antipodes. Some anglers were lured by New Zealand's sea fishing as well as its trout. Angling for kingfish (*Seriola lalandi*), a hard-fighting, streamlined species, became increasingly popular in the early twentieth century, just as the famed Taupō trout fisheries started to decline. Joseph C. Buckingham and Alexander D. Campbell, frequent visitors to New Zealand, would start their tour by targeting kingfish off Northland's east coast, setting records in the process.[18] After spending time in the Bay of Islands, they would head for Tokaanu around March in time for the latter part of the trout fishing season.[19]

Most people brought their own fishing rods and waders with them, but boats (for trolling or fishing the Tongariro delta) and local Māori attendants to serve in their camps had to be arranged locally. (For some Englishmen it was *de rigeur* to bring a valet, or at least try to hire the services of a New Zealand man.) Local businessman Captain Thomas Ryan offered a complete outfitting service for visiting sportsmen, providing tents, blankets, fishing tackle, food and all other requirements for life under canvas, right down to a camp cook. Ryan also had a small fleet of launches available to transport anglers and their equipment to the best fishing sites around the shores of the lake. This service was particularly useful for accessing rivers such as the Waihaha on Lake Taupō's western side, a region which could be approached only by water. Many groups leaving Tokaanu booked the same camping facilities for use the following season, some even leaving their fishing tackle at Taupō.

For anglers, travelling to Tokaanu at that time was a major undertaking. With very few people owning a motor car (the first one

The first roads in the Lake Taupō district were unsealed. The loose pumice became impassable in heavy rain.

Auckland Libraries Heritage Collections 35-R2241

reached Tokaanu in 1904), weekend fishing trips were impractical except for people living in nearby centres like Taupō or Taumarunui.

For those from further afield, the Taupō region could be reached from several directions. The easiest way was to take a train from Auckland and travel south to Rotorua, the heart of the thermal district. The final 60 or so miles were then covered in a coach. A second route was by road from Napier in the east. Visitors approaching from Wellington in the south could cover most of the distance by rail. The last leg, ending at Tokaanu, was a bumpy 44-mile coach journey skirting the mountains of the central plateau. But once the travellers had arrived at the angling Valhalla, there was no doubt that the effort was worth it.

Trout fishing was initially centred on Tokaanu, a tiny settlement of about 55 people in a bay on the Tongariro River delta. In 1880 a traveller noted there were some 20 Māori whares (houses), a water mill and a frame house to accommodate visitors.[20]

Being situated on low ground and surrounded by swamps, travellers did not find its scenery especially pleasant,[21] but there was accommodation at the Tokaanu Hotel, which had an attractive garden with mature fruit trees. The area also had natural features which made it attractive as a settlement to Māori. The Tokaanu Stream wound its way across the flat land to the lake, thermal vents allowed cooking and bathing, and the alluvial soils on the river delta were well suited to growing sweet potato and other vegetables. Tracks through the scrub surrounding the Tokaanu settlement gave access to the lake shore and the Tongariro River. Wild game abounded here, and anyone on foot was liable to encounter wild pigs and ducks.

Dealing with surplus catch

Not only were the Tongariro rainbows large, but, when the spawning run was on, they were numerous. This probably explains the attitude common to almost all anglers at that time of killing all fish landed, even in the absence of fish-smoking facilities. Some fish were kept for the table no doubt, but, because no one wanted a constant, unchanging fish diet, most were left for local residents to eat or were admired briefly and then buried, sometimes as garden manure.

A newspaper noted:

> Tons of fish, admittedly the finest trout in the world, have been buried on the shores of Lake Taupo in order to get rid of it; left to the gulls and other birds, . . . or have been taken off the hook and flung back into the water again.[22]

The waste of such magnificent fish, which were comparable to salmon in size and taste, did not go unnoticed. In 1910 the New Zealand tourism department began to consider ways of dealing with the surfeit of trout caught. The problem of surplus trout was not confined to Lake Taupō, either. The Lake Rotorua fishery, to the north, was already showing signs of a population imbalance, with thin, underfed specimens popularly described as 'razor fish' becoming common. The trout had outgrown their food supply. It would take government intervention in the form of a netting programme before fish numbers were reduced to a level that would improve their condition and restore the Taupō trout fishery's reputation in a second golden era.

CHAPTER 4:

The first Tongariro anglers

English visitors have been known to secure documentary evidence as to the weight of their day's baskets, . . . they knew that if they retailed their fishing experiences among their fishing friends at Home their relation of cold facts would be looked upon as fishermen's yarns.

— ***THE NEW ZEALAND HERALD*, 7 FEBRUARY 1908**

THE GIANT TROUT OF THE TONGARIRO RIVER attracted anglers from all over the world. Those who fished in the first golden era before World War I were a varied group: military officers from distant parts of the British Empire, some on leave, others pensioned off; managers of tea and rubber estates in Malaya or other colonies; and retired businessmen travelling abroad for their health. A few New Zealanders and Australians added to the mix. Apart from their love of trout fishing, the one thing these diehard anglers had in common was the time in which to pursue the sport.

Patrick C. Smith is a good example of this sort of angler. As the owner of a large *estancia* (ranch) in Argentina, he had a substantial income which allowed him to travel the world with his wife, hunting all manner of big game — tigers, elephants, bears and even hyenas all fell to his gun. He also fished for trout and salmon in remote places like Lapland. When Smith fished at Rotorua in early 1905, catching 1,234lb of trout in six weeks, it was enough to make him declare that New Zealand was truly the finest trout fishing country in the world.[1]

The magnificent Taupō fishing was available to overseas visitors at what to them seemed like an incredibly low cost. When Patrick C. Smith fished in Lake Rotorua in 1905 he concluded that New Zealand had the world's best trout, but the cost of a fishing licence was absurdly low when compared with charges in countries like Norway.[2] A few years later a visiting English angler calculated that the overall cost of the round trip to New Zealand via first-class passage, combined with associated accommodation costs, was cheaper than renting a salmon fishing beat in Scotland or Ireland.[3]

Ernest de Lautour may have been the first angler to take trout from the Tongariro River, in the late nineteenth century. De Lautour was an Englishman who had worked as a curator at fish hatcheries in South Africa, before chronic asthma forced him to emigrate to New Zealand where he thought the climate might be better for him. In 1898, while living at Tokaanu, he met Malcolm and Forrestina Ross, the travellers who had liberated juvenile trout in the Upper Waikato River.

Soon after meeting the Rosses, de Lautour returned to England, perhaps to settle some personal affairs such as a divorce (he was twice married), returning to New Zealand in 1900 to become curator of a trout hatchery at Masterton. It seems that the fisheries appointment did not last long, however, for he resigned the post and moved to Taupō in 1901 to be near his beloved trout.[4] He lived simply on the banks of the Waitahanui River, a few kilometres south of Taupō town, where his home was a raupō hut and his rent was paid to local Māori in the form of flour. It was here that he caught rainbow trout of double-figure weight; however, he kept the existence of these monsters a secret for as long as he could.

In 1906 he and Robert Jones made an impressive catch of Taupō rainbows weighing from 9½ to 20lb, some of which were sent for exhibition in Christchurch. From time to time he met visiting overseas anglers, Alfred H. Chaytor being one, and kept an eye on issues affecting the fishery by subscribing to the *Fishing Gazette*. Occasionally he would write to newspapers about such angling matters as fishing licence fees, trout acclimatisation or wildlife generally.

He must have been a trout fisher through and through, for in the 1905/06 Bay of Plenty electoral roll he gave his occupation as 'angler'. With the passage of time the asthma caught up with him, and in 1917 he gave up trout fishing for good. He died the following year, having given his name to not one but two pools on Taupō rivers — the

Waitahanui and the Tongariro. No other New Zealand angler has such an enduring legacy.

A. G. Campbell — brown trout specialist

One man who specialised in catching the massive brown trout that inhabited the Tongariro was Archibald Gowan Campbell, a very experienced salmon angler. Campbell was born in Scotland in 1868, the son of Sir George Campbell, a Scottish Liberal Party politician. He was no stranger to catching large fish even back in the Old Country, in 1906 being recorded as having taken six salmon in one day from the Hampshire Avon, the heaviest being 37lb.[5] Like Patrick Smith, Campbell must have been a man of means, because he travelled widely, seeking salmon in Canada and Scandinavia and tarpon in Florida. He seems to have been a hardy and inquisitive type, touring northern Japan in 1899 to study the Ainu, an indigenous people whose traditional life and customs were even then disappearing as the Japanese government took their land and tried to assimilate them into wider Japanese society. He also took his fly rod and successfully fished for salmon in the rivers of Hokkaido. On a subsequent trip to Norway, he spent several weeks cycling happily from one trout river to another, putting up at the local inns and staunchly fishing without waders.[6]

Campbell considered the Tongariro to be the finest brown trout river in the world. One catch of 22 fish made in April 1907 (taken at night) weighed 230lb in total. To prove to friends at home that tales of New Zealand's giant trout were not an exaggeration, he photographed his catches, placing a 2-foot ruler beside them.[7]

Fishing for large brown trout after dark in the wild, rushing Tongariro could be a dangerous undertaking. In some pools the riverbed would be made up of nothing but well-rounded stones on which the unwary could easily slip. Undaunted, the courageous Campbell would venture into the deep pools to the very limit his waders allowed, on one occasion losing his footing and almost being swept downstream. He escaped disaster only by ramming the butt of his fly rod into the river bottom to anchor himself.[8] Sometimes he would fish all night, returning with a bag of large trout as the sun rose.

Campbell counselled that the practice of night fishing was an acquired taste, and not for everyone:

> Until you get instinctively to know exactly what you are doing in the dark it is perfect misery. You get away in the bush, and you never know if you are going to find your way back again.[9]

For the first few days of a trip he would rely on the packhorse he had hired to lead him home with his catch. Most of his fish were caught on spinning tackle, but one of his biggest trout, a 23¼lb specimen, was taken on the fly — with a pattern Campbell described as a 'feathered monstrosity' bigger than any salmon fly.[10]

Campbell took many of his trout on spinning tackle, using a Phantom minnow as a lure. He was a highly observant angler and not afraid to experiment. After studying the feeding habits of the Tongariro trout, he fashioned a minnow using the dark-coloured cocoon of a moth which he found hanging on the riverside mānuka trees. Results proved that the cocoon was just the right shade for imitating the kōaro and other small fish, although it had to be varied in its size and colour, and the depth at which it was fished. So well-attuned did Campbell become to the habits of the Tongariro browns that when approaching a pool at night he could instinctively sense from weather and water conditions just what technique was required for success. A contemporary who fished with him on the Tongariro considered him the best angler he had ever seen.

In early March 1907, Campbell told W. H. Tisdall, the New Zealand fishing tackle dealer, that he had discovered that the Tokaanu trout would take the fly at night.[11] Campbell had his two largest trout, weighing 23lb and 23½lb, mounted, and sent them to London, where they were exhibited in the shop window of Hardy's, the well-known rod makers, as proof that New Zealand was the 'angler's Paradise'. Like Joseph Buckingham, whom we shall meet shortly, Campbell found the Tongariro fishing not merely irresistible, but practically indescribable.

The Field described Campbell's trout with awe:

> . . . as fine a brace of trout as is ever likely to be seen in this country. . . . Both skins are very silvery and somewhat profusely spotted with small dark spots. The heads are relatively small, and the fish give the impression of tremendous girth and thickness. They are a great tribute to the food supply of Lake Taupo, whence they got into the river.[12]

Archibald G. Campbell, rod at his feet and two Tongariro River brown trout in hand weighing 23¼lb and 17lb. The fish were caught with minnow and fly on 1 March 1907.
The New Zealand Mail, Issue 1833, 24 April 1907, Page 4 (Supplement)

Campbell later explained to Tisdall how he had first used flies on Tongariro trout in darkness:

> It is a most satisfactory discovery that the big lake fish will take a fly. A strong hook is necessary, but I'm inclined to think the smaller the better, as, especially in fly-fishing, the fish take so gently and eject

> the lure so quickly, that the great difficulty is to get the hook into their tough jaws over the barb.
>
> . . .
>
> My biggest fish this time have been a couple of 23-pounders, one of 21½lb and eight between 17lb and 20lb in weight. . . . those who have been accustomed to salmon-fishing at Home have been able to catch fish here, and I fancy there will be some very good takes with the fly when it becomes known that this is really an effective way of fishing.[13]

Campbell got his best catches when the moon was out, taking half a ton of trout (100 in total averaging more than 11lb each) in 12 days — or, more accurately, nights. On a cloudy night with no moon there was little action.

The fighting qualities of the browns would vary; often it would be a long, drawn-out struggle with the fish imitating a log. Only in the hour before daybreak would they became much more lively: '. . . all of a sudden they appear to wake up and to vie with one another in giving splendid exhibitions of wild rushes and jumps, and on other nights there will not be a dull fish amongst them.'[14]

Campbell returned to England, where more large salmon awaited him. On 4 May 1916 he caught two massive fish, weighing 33lb and 43lb, at the site of his triumph a decade before, the Hampshire Avon, at Ringwood. It appears such extraordinary angling feats earned him a reputation as an expert, because the Scottish tackle dealers P. D. Malloch produced the 'A. G. Campbell salmon spinning rod', which they advertised in their 1933 catalogue. He died in London in 1957.

W. S. Pillans

Another man who livened up the Tongariro scene was William Soltan Pillans, a retired farmer lured to the thermal regions by tales of big fish. He subsequently spent the first decade of the twentieth century fishing the river and Lake Taupō over the summer months, retreating each winter (the closed season) to the warmer northern districts. When floods or bad weather made fishing impossible, he entertained other residents at Hut Camp from a store of anecdotes, or carved remarkably life-like trout from wood to give to others. A photograph taken at the time shows him holding two trout, one of 18lb, the other a pound lighter.[15] He died in 1915, shortly after the Taupō fishery began its decline.

A. D. Shilson

Arthur Dinham Shilson was born in Devon, but came to New Zealand as a young man. In the 1890s he began farming at Waingaro in the Waikato, to the north of Lake Taupō, and first came to fish the Taupō waters around 1904. Accounts of his exploits may have attracted Major Rhys Jones to emulate him.

Shilson had a penchant for catching his limit, and he liked to let others know about it. For him, trout fishing was a numbers game; to use the old golfing adage, it was not how, it was how many. Over the 1909–10 fishing season he banked 1,248 Tongariro trout, at a total weight of 9,839lb and an average weight 12.6lb.[16] His heaviest trout weighed 18lb.

Records kept by Captain Thomas Ryan, the fishing camp outfitter, show that Shilson was a long way ahead of his nearest rival for the 1909–10 season, Colonel Moore, who caught 566 fish from the Western Bay rivers. Both men caught an 18lb trout. The following season was even better for Shilson: 1,611 trout, with a total weight of 14,771lb and an average weight exceeding 9lb.[17] His biggest bag for one day was 42 fish with a total weight of 419lb — how his arm must have ached under the strain of a long split-cane rod with a massive rainbow trout on his line in such heavy water!

Shilson courted publicity to some extent, even writing to local newspapers with details of his impressive catches and claiming the Taupō fishery to be the world's best, far surpassing the exclusive salmon angling of his English friends.[18] Shilson, who had fished in England and Scotland, considered that for brown or rainbow trout, no other country could top New Zealand.[19] The Tongariro River, with its fast-flowing waters, was his particular favourite because the fish fought so hard, and in that respect equalled salmon of the same size. He also pointed out that the rainbows were being caught on the fly.

According to Malcolm Ross, who met him in those early days, Shilson often fished from a boat in the lower reaches of the river. He seems to have been something of a lone wolf, being totally focused on the pursuit of more and bigger trout. Ross saw one example of this on the Tongariro:

> . . . Archie Clark had got out of his depth and had to swim for it in the big pool opposite our camp. Shilson was close at hand in his boat, but

Arthur Dinham Shilson, a fixture on the Tongariro in the early days. His catch for one day's fishing totalled 18 trout.

Auckland Libraries Heritage Collections AWNS-19110323-7-3

Arthur Dinham Shilson, left, with two fine Taupō rainbows. Small wonder the newspapers called them 'salmon trout'.

Right, two fishermen preparing a meal on the banks of the Tongariro River.

Auckland Libraries Heritage Collections AWNS-19110420-3-1

calmly went on fishing. Finally, when Archie struggled out of the pool — and it is not easy to do that with long waders on and a rod in your hand — Shilson called from the boat: 'Did you get wet?' At the moment Clark could think of no suitable reply, but he said a great deal afterwards.[20]

By 1910 Shilson was voicing the opinion that the Taupō fishery had so many trout they needed thinning out before their condition deteriorated. That may explain why he could never catch too many. Like Major Jones he was a sunseeker, spending the southern winter in the islands of the South Pacific; in Shilson's case it was Tahiti, where he was a plantation owner.

Shilson died in New Zealand in 1924, and was buried in Auckland's Hillsborough Cemetery. Like many of his contemporaries who made fly fishing their first priority, Shilson left no wife or other dependants.

The almost blasé attitude of the anglers to the enormous bags of trout they took from the Taupō fishery was summed up by Arthur Millington Naylor, an Englishman who, without having to fish very hard, 'was content with an aggregate of two ton of fish for the season'.[21]

Major R. W. W. Jones

One of the first anglers to explore the Tongariro river lent his name to one of its most popular and prolific pools. The Major Jones pool, in the river's middle reaches, is named after Major Rhys William Wykeham Jones, a Welshman who made regular visits to New Zealand before and after World War I. He is thought to be the angler to whom O. S. Hintz, author of *Trout at Taupo*, refers as using field mice to catch large brown trout at night. Jones considered these fish cannibals a menace to the fishery. In 1909 he was onc of thc first anglers to seek out the pools above the Tongariro delta, using flies instead of spinning tackle.[22] Joe Frost's description portrays him as a giant of a man, more than 6 feet tall and large in the girth, sporting a handlebar moustache and a monocle.[23] A man of that size would have had no problem wielding a 17-foot fly rod on the Tongariro's bigger pools.

Jones was born on 27 April 1863 in Loughor, near Swansea in South Wales. The son of a solicitor, he attended Bath College where his sporting pursuits included rugby and cricket, two classic pastimes for Victorian youth. Of fishing there is no mention, so it is most likely he took up the gentle art when posted to India in the 1880s.[24] Tragedy occurred early in his life when his mother died in 1864, the year after he was born. His father died one year later, leaving the infant Jones and his younger brother as orphans. Some parish organisation must have stepped in to help, because by the age of eight he and his brother were boarding with the Surgeon-General, Hugh Massey in Bath.[25]

Lt R.W.W. Jones, 1st Battalion, Royal Sussex regiment, Portsmouth 1885.
West Sussex Record Office, Chichester, England.

In February 1882, Jones entered the Royal Military Academy Sandhurst, just two months before his nineteenth birthday. After one year's training Gentleman Cadet Jones passed the course, but graduated ninety-ninth in a class of 105 men, which suggests he was not the finest officer material. He received a commission on 10 March 1883[26] and joined the 35th (Royal Sussex) Regiment in its expedition to the Sudan to relieve the besieged General Gordon in his fortress at Khartoum — an imperial errand that proved to be a masterpiece of mismanagement and procrastination.[27] While on the Nile, Lieutenant Jones fell ill, and returned to Cairo on 10 June 1885. He was invalided and left Egypt for England, arriving there in October 1885.[28]

Jones was promoted to captain on 13 April 1892, and retired several months later, on 26 October 1892,[29] although he later took part in the Boer War with the 3rd Border Regiment. He left South Africa on 18 September 1901, and was back in England the following month.[30]Although he resigned his commission he was awarded a post-dated commission and pension for militia work. He kept permission to use his rank, and from that day forth he was known as Major Jones.[31]

Jones then began to travel extensively; so much so that according to the records of the Bath Club, he lived mostly outside the United Kingdom from 1895 to 1908.[32] In that period he returned briefly to England for one year, before arriving in New Zealand on 13 March 1907.[33] The South Island attracted him initially, but after two months he moved north towards the central North Island where he spent the rest of the year. This 1907 tour was very likely his first visit to New Zealand. In an interview he gave to a Wellington newspaper in December 1907,[34] he commented on the price of land, the reality of life in New Zealand compared to the picture painted in his tourist guide, the need to hire a motor car to get around, and the unskilled methods of local anglers. All these views indicate someone new in the country and getting a first impression. He also visited the New Zealand Government Tourist Office in May and December 1907, probably for advice on when and where to fish.[35]

The Major had no cause for complaint about the quality of the fishing, for in March 1908 he caught a 17lb rainbow on the fly from the Tongariro River.[36] The fish was sent back to the United Kingdom, preserved in ice, as proof that Taupō trout were indeed as large as they were said to be.

New Zealand became his favourite fishing destination. His name appeared periodically in the society columns of daily newspapers reporting on guests staying at the best hotels in Auckland and other main cities. Records indicate he was present in 1907,[37] 1909[38] and 1911,[39] and by 1912 newspapers were referring to him as a 'well known English angler'.[40]

Major Jones was one of the first anglers to fish for Tongariro rainbows with an artificial fly. (A. G. Campbell, as discussed earlier, had already taken brown trout with salmon flies, but he fished at night.) John Bushby, writing in *The Field* in 1910, records that Jones was among the party which camped 4 miles up the river, their purpose being 'strictly for fly-fishing', and Jones was really the first successful angler to use that method on the river, instead of a spoon or a minnow.[41]

Bushby's camping trip took place in 1909. The group explored the Tongariro River upstream from the Taupō highway bridge near the Island Pool, and the deep, slow run above it, which would come to be called the Major Jones Pool.[42] It is unclear precisely when the Major Jones Pool gained its name. Newspaper records do not refer

Major R.W.W. Jones wearing medals for active service on behalf of the British Crown.
Target Taupō January 2012, Issue 64, p 47

to it as such until 1924, some two years after Jones died.[43] Perhaps the name was conferred after his death as a mark of respect to the man who discovered the opportunities for fly fishing on the river's middle reaches.

Another visit Major Jones made in early 1912 was marred by illness. After recuperating in Auckland, he returned to the United Kingdom in April 1912, before arriving on another fishing expedition in December of that same year.[44] He left the United Kingdom only twice more, one journey being to New Zealand in 1921.[45] For Jones and the rest of the Tongariro regulars, it was their practice to remain in New Zealand to fish most of the season and then depart for warmer climes.

Jones was typical of many Englishmen who travelled the Empire in the early twentieth century to hunt, fish and explore. Most of them were from the upper classes and had private incomes, which meant they had no need to work. In those days, men from the more privileged families either went into the professions (usually medicine or law) or the armed forces. If none of these paths were taken, they were simply listed as 'gentlemen' on shipping passenger lists. The colonies were their playground.

Jones does not seem to have been obsessive about his sport. That characteristic made him an ideal angling companion; never competitive or superior in angling knowledge, and as happy to see another catching a trout as when banking one himself. A friend who fished with him in England said that so long as the major was beside a river, he was happy. On occasions he would accompany another angler just to watch, taking nothing more than his pipe.

Major Jones died on 2 August 1922 at Bath, aged 59 years. Sadly, his death came on the eve of a trip to New Zealand for what must undoubtedly have been another expedition to the Tongariro. He had booked his passage and sent his rods and other equipment on ahead, but was himself delayed by the need for what was described as a slight operation. Shortly after, pneumonia set in unexpectedly.[46] His headstone in Bath cemetery records him as 'mourned by friends all the world over'.

Major Jones Pool, April 1951.

Joyce Galbraith. Supplied by Taupō Museum and Art Gallery.

The Major Jones Pool is one of the most popular pools on the river. It is visually appealing, with a small gravel beach at the head where anglers enter the water. Opposite, a large tree overhangs the water on the high bank which runs the length of the pool. After the rapid at the head, the pool widens and the river slows to a quiet pace.

Even after the Tongariro Power Development scheme reduced the river's flow, the pool retains its attractions both for trout and angler alike. For the fish, the slower-flowing water at the tail, coming immediately after the rapid at the head of the Island Pool, gives welcome rest. For anglers, there is no white water to make wading difficult, the bottom is fine silt or sand, and casting to the far bank is easily accomplished. The real challenge is to get the fly to swim deeply enough.

Despite the effects of several massive floods, the Major Jones Pool has survived as a living monument to one of the first great Tongariro River fly fishers. Other pools may come and go, but it will surely remain a firm favourite with many anglers.

H. C. King-Webster

Harold Colin Webster, a Liverpool solicitor, was another member of the first Tongariro brigade. He fished with John Bushby and Alexander D. Campbell at Tokaanu in January 1909, where he caught a 17½lb rainbow at the Tongariro River mouth.[47] By the time he returned to fish Taupō waters in February 1912, he had married and changed his name to King-Webster, so perhaps this latter trip was a part of his honeymoon.

When World War I broke out King-Webster rushed to join the colours, even though he was 35 years old. He enlisted in the 23rd (Sportsman's) Battalion of the Royal Fusiliers on a short service commission. This volunteer unit took men up to the age of 45 years who, because of their active lifestyle, were thought to be fit and strong enough for military service. Unlike the 'pals' battalions formed in the early days of the war from towns like Sheffield and Hull, the 23rd drew recruits from all over the country and all walks of life. One man, the brother of a peer, enlisted along with his chauffeur.

King-Webster survived the conflict and returned to his manor at Wanwood Moor in Cumberland. He maintained an interest in New Zealand, however, and in 1923 sent grouse from his estate as a gift in an attempt to establish that species in this country.

The English angler John Bushby fished at Tokaanu in early 1909 with Harold King-Webster (who was then called simply Webster):

> . . . we took coach, fifty-six miles, to Lake Taupo, crossing to Tokaanu, where we had arranged with the hotel proprietor to put us up a camp on the Tongariro river. Needless to say, and luckily as it turned out for us, the camp was not ready, and we determined to give the mouth of the Tongariro where it enters Lake Taupo a good trial with the fly. So far as I can ascertain, very few, if any, fish had been taken hitherto, so that we may perhaps consider ourselves the pioneers of successful fly fishing in the lake.
>
> . . .
>
> . . . the river near its outlet is surrounded by swamps and can only be approached by boat, it is isolated except in fine weather. Weather being fairly favourable, we rowed across the bay a mile to the first outlet, where I was dropped to try my luck, whilst the others proceeded a mile to the main channel. My efforts resulted in nothing, and about midday I moved on along the shore to join the others. Here I found a large fish had been

> moved, which gave us hopes, and it was not long before W. was fast into a monster which threatened to take rod and line away with him. As we were all fishing with 11 ft. rods, this was exciting, and I stood by to give what assistance I could. Careful handling at last brought the first fish into shallow water, and it was duly gaffed. Truly a magnificent fish, 17½ lb., measuring 31½in. long, 21½ in. greatest girth, as silvery as a fresh-run salmon. Shortly after this P. was fast in a fish some hundred yards away, and after exciting runs and jumps landed an 8½ lb. fish. Again W. was lucky enough to get an 11½-pounder, whilst I had to be content with expectations, for never a rise rewarded me the whole day.
>
> This was good enough, however, — in fact, it seemed a land of promise — so that, having selected a lovely place for a camp on an island in the river itself, we returned home laden with fish, on the way discussing the probabilities of much larger fish to be caught.
>
> . . .
>
> My own bad luck was broken the second day out, when I landed three fish, of 9lb., 9½ lb., and 11lb., within half an hour. To keep the find quiet was impossible, and it was not long before we had other fishermen trying for a share of our good luck. Had they confined themselves to fly we could not have found much fault, but it was a case of trolling backwards and forwards across the stream and disturbing the water with large spoons. . . .
>
> Although you see the rainbow jumping and breaking the surface, it is seldom that they will take a fly actually on the surface in the lake, and our method of fishing for these huge trout was very different from that followed for ordinary trout. It is on the inanga or whitebait that they are feeding, therefore it is essential to fish with a fly sunken more or less, casting a long line, and very gradually working it in against the current. The most telling flies we found to be Silver Doctor, Wilkinson, small to medium grilse size. Stout casts and 100 yards of line are essential. Even with this to be clean run out has happened more than once.

The group stayed at Tokaanu from 7 January to 4 February 1909, taking some 110 trout, at an average weight of 10lb. These were rainbow trout; fishing for brown trout required a different method, as Bushby noted:

> The Tongariro hitherto has been considered only good for brown trout, which run to an enormous size, but will only feed at night. A walk down

> the river on a bright day is a sufficient recompense to anyone interested in fishing. In certain bends of the river we used to imagine that the water actually smelt and tasted fishy. At any rate, within 200 yards I have seen that number of brown trout, varying from 8lb. to 20lb. in weight, basking on the shallows and only slowly moving off into the current when disturbed. All manner of lures have been tried to tempt them during the day, but it is useless, and it is only as night approaches that you have a reasonable chance. Men make Tokaanu their headquarters, fishing all night and sleeping all day. Enormous catches are sometimes made, generally on a still moonlight night with passing clouds.[48]

J. W. Moore

Scotsman Colonel John Wardrop Moore was another who visited New Zealand to hunt and fish. He had inherited his family estate at Blantyre in South Lanarkshire in 1885, which allowed him to live the life of a gentleman farmer with time off for outdoor pursuits in foreign parts. He was to make six annual visits to New Zealand for trout fishing, the last one in 1912. In 1911 he took 3 tons of trout from Taupō, which perhaps explains why he thought it was 'the most wonderful place on earth for trout'.[49]

Many of his friends in the United Kingdom refused to believe the tales about 20lb trout from Taupō, so Moore had several specimens mounted, and returned home with the trophies as indisputable proof.

He died on 16 October 1912, aged 49 years. His early death was thought to be caused by injuries sustained in New Zealand two years earlier, when he fell 300 feet down a cliff while stalking a deer. He broke several ribs in the fall, but that was not the end of his ordeal. Friends carried him to the nearest road from where it was a painful ride to town on a buckboard wagon. However, on the way the horses bolted, throwing him off and into a stream bed.

Moore spent six months in a Dunedin hospital recovering, but emerged as keen as ever to cast a line. When doctors warned him that it would probably be fatal for him to fish again that season, the irrepressible Moore replied 'Well, if I'm to die, I'll die among the fish', and promptly went to Taupō, where he caught more than a ton of trout.[50]

Joseph C. Buckingham

One man, Joseph Charles Buckingham, had the good fortune to experience the two main growth periods in the Taupō trout fishery, which came at the start of the twentieth century and in the early 1920s. Born into a family of London silk merchants on 18 June 1864, Buckingham attended Charterhouse school before gaining a degree at Trinity College, Cambridge, in the 1880s. His fishing began in Yorkshire, where the trout were weighed in ounces, but all that was to change.[51]

Together with his brother Henry, Joseph Buckingham joined his father's London firm in 1886 as a partner, but does not seem to have been overly attracted to business life. Once his travelling started, there is no evidence he felt the need to work at all. Leaving others to run the firm (Henry would become managing director in 1899),[52] Joseph undertook the first of many voyages to New Zealand.

His first New Zealand visit took place in 1886 when he was only 21 years old. At the time, New Zealand was a raw, undeveloped land and conflict with indigenous Māori was in the recent past. On his first visit, Buckingham arrived in Rotorua by wagon, shortly after the Mt Tarawera eruption which had devastated the nearby landscape and claimed the lives of local residents. That calamity must have underscored the wild, unpredictable nature of the country, but failed to curb Buckingham's enthusiasm to explore.

In 1898 Joseph Buckingham inherited a share in his late father's estate, worth £168,138 (some £16.3 million in today's money). The door to a life of sporting leisure now stood wide open, and New Zealand trout fishing, along with cricket, was to be a major part of his life. Perhaps not surprisingly, he never married.

The size of Taupō's trout was a revelation to him. In 1908 he was one of the first anglers to fish at the Waitahanui River, a Lake Taupō tributary to the north of the Tongariro, where he found himself catching trout 'I had never anticipated being really permanently connected with on this earth . . .'[53] The other feature of trout fishing in New Zealand which Buckingham had not anticipated was the need to guard his catch against the marauding pigs owned by local people, and which had developed a taste for rainbow trout.

Not even a global conflict stopped him travelling to enjoy his sport. Buckingham's name appears on the passenger list of a ship sailing from London for Auckland in late October 1915. (At that stage, Germany had declared a war zone around the United Kingdom, but did not begin unrestricted submarine warfare until February 1917.) Perhaps wisely, he planned to remain in Australia for several months after that. His travel papers described him as a 'gentleman' or 'retired'. He visited in 1916 and 1917 also,[54] but found time for national service in World War I, applying his skills as a linguist and acting as an interpreter for the British Army.

Like his companion Shilson, Joseph Buckingham believed in culling the trout population. At the end of World War I, when speaking of the golden days of big trout, he said:

> As regards the enormous bags of heavy fish made in Taupo Moana eight and nine years ago, when our average was 10lb, per fish for many hundred fish, the painful word is Ichabod; . . . in those non-netting days, the more fish killed, even though wasted, the better.[55]

By the end of the 1926–27 season, perhaps earlier, Buckingham's diary recorded that he had taken 4,092 trout at a total weight of 23,493lb. These amounts were conservative, because he measured trout in total pounds only, ignoring the ounces.[56] Most of the fish were caught on artificial minnow, but some fell to the lure of a large salmon fly such as a Durham Ranger, to which a lead weight was added (this Buckingham described as 'worming with the fly').[57] His heaviest trout, landed in 1907, weighed in at 19lb, while his average weight of fish taken the following season was 10lb.[58]

He had a decided preference for outdoor action in Australia and New Zealand rather than the south of France and other 'tourist-infested spots'. So frequent were his absences from England that he usually gave The Sports Club, St James, London SW as his home address on immigration forms. Buckingham became a confirmed Tongariro regular, and completed his twenty-ninth visit to New Zealand's shores in March 1928. By that time, he and Alexander Campbell were the sole survivors of the old Tongariro brigade. With the rigours of old age upon him, he was unable to wade the river and cast in the usual style, so he switched to fishing from a boat at the river mouth. Despite these limitations his enthusiasm was undimmed, even though the days of catching 50 trout in a day were long gone.

In November 1929 he arrived in New Zealand on his thirty-first visit.[59] Several years later, in 1933, he estimated he had travelled 1 million miles by sea in pursuit of trout and big game fish.

He died in Ramsgate, England, on 2 June 1939, aged 74 years,

Alexander D. Campbell, pioneer of New Zealand deep-sea fishing, with a striped marlin taken in 1915.
Auckland Libraries Heritage Collections AWNS-19150218-50-4

leaving an estate valued at £112,303. His place of residence at that time was given as Brents Hotel, Rotorua.

Alexander D. Campbell

Alexander Drummond Campbell was a Scots angler who came to New Zealand in the years before World War I. A tea planter from Ceylon (now Sri Lanka), he first visited New Zealand for health reasons in

1900. At that time he was a young man of 23 years. An Auckland doctor, James Moir, treated him for malaria, and when the two men discovered they had a common interest in sea fishing they began a lifelong friendship.[60]

Although keen on fly fishing, Alexander Campbell also helped to develop tackle for use in big game fishing in the Bay of Islands.[61] In 1914 he was fishing from a boat there for medium game fish when a marlin seized the bait and broke the line, giving him a glimpse of the enormous power and grace of these billfish. Fired by the possibilities of landing one, Campbell set about designing tackle strong enough for the task. When he returned the following year he was successful, and is thought to be the first person to take a marlin in New Zealand waters on rod and line.[62] It was New Zealand's sharks, marlin and swordfish which were later to play a part in luring Zane Grey to Antipodean waters.

Campbell was so keen on New Zealand's fishing that he returned almost every year. By 1933 he was making his twenty-third visit, undeterred by poor health in the form of a heart problem.[63] He died that year with a rod in his hand, fishing the Waikato River near Alan Pye's Huka Lodge.[64] Like his fellow anglers Shilson and Major Jones, he was a confirmed bachelor.

Charles E. Lucas and Gerald N. Horlick

As noted in Chapter 3, Charles Eric Lucas ('Joe' to family and friends) was one of two English anglers who caught some of the Tongariro's massive rainbows in 1912. Lucas was accompanied by Gerald N. Horlick, a member of the family famed for producing Horlick's malted milk. They stayed on the river for two weeks, using fly and spoon to bag 197 trout, only three of which were browns. On 13 March, Charles Lucas caught the largest trout at Gull Beach, a fine 20lb rainbow which rose at a Durham Ranger fly.

The pair then moved to the South Island for some deer stalking, Lucas getting a 12-pointer before returning to England. When World War I broke out, Gerald Horlick was with the Royal Gloucestershire Hussars and saw service at Gallipoli and Egypt. He proved to be a very likeable member of his unit; popular with both fellow officers and the men under his command. Major Gerald Horlick came through a number of battles unscathed, only to be felled by malaria in Alexandria in 1918.

Australian and New Zealand anglers

Many New Zealand anglers fished the Tongariro River, but few can have been as constant in their efforts as Lionel Hanlon, who fished at Taupō for decades. Irish-born, Hanlon came to New Zealand in 1881 as a young man and started fruit farming in Northland. He excelled at horticulture, experimenting with new fruit varieties; so much so that in 1900 he represented New Zealand in France at an international fruit exhibition. The plums he showed there are thought to be the first of their kind shipped from New Zealand to Europe.

Hanlon fished at Taupō for 40 seasons in a row — surely a record. He made his last journey there in 1945, shortly before he died at the age of 87 years.[65] A keen photographer, when Zane Grey was on his first New Zealand visit in 1926, Hanlon met Captain Laurie Mitchell, the author's friend, and passed on photographs of the fishing at Taupō. He was the first president of the Upper Waikato and Tongariro Anglers Club, and played a prominent role in the club's activities at Tūrangi, lobbying for improvements to the Taupō fishery.

Hanlon's heaviest trout was a 19lb brown he caught in the Tongariro River in 1911 on a silver Devon minnow. His total catch for two weeks' fishing was 31 trout, averaging an impressive 11lb.[66] Hanlon was one of the group of early Tongariro anglers, including William Soltan Pillans and Joseph Buckingham.

The magnificent New Zealand trout fishing did not escape the notice of Australians, who, with no rivers to compare with the mighty Tongariro, made the trip to Tokaanu to experience the world's best rainbow trout fishing. One such was Charles Potts of Sydney who fished the Lake Taupō tributaries at the start of the twentieth century. In 1905, fishing for two months, he and his wife, Blanche, caught 2½ tons of trout without huge effort, making sure to take photographs to prove their tales of New Zealand trout were true.[67]

Among the New Zealand anglers, Robert Jones reigned supreme on the Tongariro River. Together with Taupō's Thomas Ryan he was one of the most successful businessmen in the region. Jones set up several anglers' camps on the lower river. Living at the waterside the whole year round gave him the opportunity to observe the trout and learn their habits. But even allowing for this local knowledge, he had angling skills which gave him the edge. His name appears frequently in newspaper reports of Tokaanu trout catches, one of which records his bag of 15 trout weighing from 7lb to 14lb.[68]

Other Tongariro anglers marvelled at Robert Jones's expertise:

> Many a time I have watched that good fellow, after delivering supplies to the camp, come with his spinning rod to The Parade to get some fish for the house; seen him arrive at the head of the pool, and with a slight flick of his rod send the Gold Devon to within a few yards of the other shore; give a few turns of the reel and chink — a tight line. Then one would hear that screech of the reel that makes every angler thrill, see the leap of a shining beauty . . . and the securing of a 10-pounder. Back Bob would march to the head of the pool, and after waiting for a few moments, . . . while another angler tried his luck, . . . flick out again the Gold Devon with like result, this time a 12-pounder, and so on and on. He seemed absolutely to mesmerise the fish, for in the course of an hour or so he would have landed 10 to 12 beauties, while other anglers had secured one, or two each, perhaps.[69]

Early fishing techniques

Just as the trout were transplanted from the Northern Hemisphere, so, too, were the methods used to catch them. In the early days of the Tongariro fishery, when trout the size of English salmon were common, anglers used salmon rods up to 16 feet long. The flies, too, were at first based on those of the Old Country: salmon flies like Thunder and Lightning, Bulldog and Silver Doctor tied with a heavy wing. Over time, the propensity of the rainbow trout to take almost any fly became obvious. Local patterns began to emerge using plumage from New Zealand birds, for example the Mallard, Maori Chief and the Matuku (bittern). The last of these was especially effective because it imitated the colour of small bait fish.

Anglers fishing the Tongariro River in the early days had few restrictions on the methods they could use. Fly, spoon and minnow (Devon) were all allowed, but the regulations excluded natural bait such as worm.

The fast-flowing rapids and deep pools of the Tongariro River required lines to be weighted to get the lures down to the fish. Lead weight added to the line was standard practice and breached no regulation. All of these angling techniques were thrown together on the lower reaches of the Tongariro, below the main highway bridge, where most of the fishing was done.

Big trout in heavy water were not for everyone, and not every visiting angler found the lower Tongariro fishing to his liking. Canterbury angler G. F. Whiteside tried boat fishing at the delta in 1913. His catch was impressive — 38 fish in five days, at an average of 7½lb — but he thought it a lazy affair suited to old men. He found more attraction in fishing for the 3lb rainbows which rose readily in the Tokaanu stream, close to his hotel, and which could be taken at any time of the day.[70]

Tongariro tackle

What would the well-equipped Tongariro fly fisher have been equipped with in the early days, say the 1920s? They would probably have chosen a light salmon rod from a famous maker like Hardy's — perhaps the Gold Medal, a double-handed three-piece model, 14 feet long and weighing around 19oz. 'Guaranteed superior to any other make' (according to the 1923 Hardy's Anglers' Guide), it would have been fashioned from the finest 'Palakona' split-cane and sold with a spare tip.

For dry fly work, a Pope dry fly perfection rod might be selected, 10 feet long and in two pieces. Ernest Hemingway once wrote that there are no 100 per cent fly fishers, so a spinning rod would be included as well to cast a minnow.

Candidates for the fly reel included the 4-inch Hardy's Perfect, weighing 12oz and boasting a compensating check mechanism, or the St George or Uniqua models, any one of which had good line capacity and would make an excellent match for the salmon rod.

The reel for the spinning rod would almost certainly have been the Silex No. 2 casting reel, which was destined to spark the great Tongariro spoon controversy of the 1930s. Hardy's were justifiably proud of this free spool reel, which was the result of years of experimenting and field testing, and was used to set several British casting records. It featured a regulator mechanism which, once properly set to correspond to the weight of the minnow being used, would ensure overruns and tangles were a thing of the past.

Other essential items of tackle were a telescopic gaff, salmon and trout flies, chest waders matched with leather brogues studded with non-slip hobs, a rain jacket, spare reels and silk fly lines. Perhaps most important of all was the weakest link in the chain connecting angler and trout: the cast or leader made of Spanish silkworm gut. Spare casts were carried in a flat, circular cast case of japanned tin, or in a leather pouch, both having a felt damp-pad to keep the gut moist and ready for immediate use; dry gut became brittle and was liable to break.

Creels of the kind used on British trout waters were unknown on the Tongariro River, being far too small to take even a few Taupō trout from those halcyon days. The fish were either carried on a wooden-handled stringer or left on the riverbank for the staff from one of Robert Jones's camps to collect later.

Being well aware that anglers love gadgets, Hardy's offered several handy devices: a combination wading staff and gaff, a spring balance to weigh the prize, scissors, a hip flask and a magnifying lens attached to the thumb for use in attaching flies ('a boon to those with defective sight'). And no real angler would be without a priest with which to speedily dispatch their fish.

Taupō trout attracted both men and women anglers. Mrs Blanche Potts of Sydney poses with a superb Taupō rainbow, probably a double-figure specimen, caught in 1911.[71]

Auckland Libraries Heritage Collections

CHAPTER 5:

Robert Jones's private fishery

Whenever there's room for a fresh business venture in Tokaanu you can guess who'll start it.

— ***NEW ZEALAND TRUTH*, 16 FEBRUARY 1924**

SINCE THE MID-1920S, anglers venturing to Lake Taupō have enjoyed legal access on foot to the lake and the rivers which feed it. This was not always the case. Before 1926 much riverside land was held by Māori, the original inhabitants of the region, under native title. And because ownership allowed control, the Māori landowners could charge a fee to anyone who wanted to cross their property to reach the river. In the early twentieth century an astute Tokaanu businessman, Robert Jones, took advantage of this situation to strike a deal which resulted in his establishing a magnificent private trout fishery.

Robert Jones came to Tokaanu from Taihape, where he had been a successful farmer and had developed a reputation as a fine horseman. Shortly after the turn of the century he must have decided on a change of career and acquired Blake's Hotel at Tokaanu, the only licensed liquor outlet in the district. There had been a hotel at Tokaanu since 1874, when two Hawke's Bay businessmen had opened it with a modest four rooms for travellers. The year 1883 saw the hotel name changed to Blake's Hotel when Sgt-Major George Blake of the local Armed Constabulary acquired the business in association with John Grace.

The hotel was in a prime location, being the only accommodation point for tourists making their way to or from the tourist attractions of Rotorua and the scenic Whanganui River at Pipiriki. Blake held the land on which the hotel stood under a long-term lease of 21 years. When he tried to sell the business in 1902, however, there were no takers, and it was not until 1904 that Robert Jones came on the scene and took over.

This was Jones's first step in forging strong links with the liquor trade. The following year he married Annie Asher, the daughter of David Asher, a well-known Tauranga publican, and the couple settled into married life in charge of what was now called the Tokaanu Hotel. By the end of 1906 the accommodation licence for the hotel was in Jones's name; the accommodation licence being a licence to sell liquor on condition that the licensee also provided accommodation to the public at the premises.[1]

Although George Blake had tried to publicise the local trout fishing, there had been one drawback: at first the only fishing on offer was for brown trout, which, as we have seen, did not take a lure readily in daylight. Few tourists would have been keen to risk scrambling along the Tongariro's boulder beds in lonely darkness in the hope of making a catch. The arrival of the rainbow trout and the angling skills of Robert Jones changed all that.

With the hotel and liquor trade firmly in his grasp, Jones proceeded to build up what amounted to a private trout fishery on the banks of the Tongariro River. A clause in the lease he had signed with the Māori landowners gave him exclusive occupation rights for riverside land. Anyone who wanted to fish the lower pools below the main highway bridge had to stay at the hotel or at one of Jones's camps.

The riverside camps

Initially, Jones developed a rudimentary trio of camps for anglers. One was on 'Jones Island' at the mouth of the Tongariro delta, on the true right bank of the Tongariro's main stem, some 400 metres from the river mouth. This was Delta Camp. The other two, a little way up the river, bordered the Jones Pool and the legendary Hut Pool, respectively. (The last so-called because of the 10 huts Jones had built for anglers' accommodation.) It was known as the Hut Camp.[2]

In the first decade of the twentieth century, it was the Hut Pool which drew anglers from across the world. It was easily accessible,

The lower Tongariro River, looking towards Lake Taupō, 6 April 1949. The Hut Camp buildings are on the true left bank in the centre of the photograph, adjacent to the rapids at the head of the Hut Pool.

National Library image WA-19984-F

Hut Camp buildings. In the background at the right of the picture trout are drying, possibly in preparation for the smokehouse.

National Library of New Zealand. White's Aviation 28351-017594.tif

The finest trout fishing country in the world: An angler landing a 12-pounder on the Tongariro River, Lake Taupō.
Auckland Libraries Heritage Collections AWNS-19130501-10-2

accommodated several rods, and yielded great numbers of large trout season after season. Many Tongariro devotees grew to like it so much that they fished no other water.

The Hut Camp location was a short distance below the Tūrangi road bridge. From that point the thick bankside vegetation (mānuka trees) made it difficult to fish the water further upstream.

Fishing camp life — the early days

Once established at one of Jones's camps, the visiting anglers had to study the techniques of the locals, adapt to the conditions, and try for trout as best they could. For the English anglers it was trout fishing, but the size of the fish required salmon tackle. The guests were a varied bunch; titled English lords and ladies touring the dominions, and British Army officers on leave from far corners of the Empire,

TOP
In quest of New Zealand's giant trout: an anglers' camp on the banks of the Tongariro River, Lake Taupō.
Auckland Libraries Heritage Collections AWNS-19110330-14-6

BELOW
A successful group of anglers at the Hut Camp, November 1910. Note the long fly rods in use.
Auckland Libraries Heritage Collections AWNS-19101110-3-1

would mingle with Australian tourists and a few New Zealanders. In early 1910 even the Governor-General Lord Plunket, was a guest.[3] But when it came to catching trout, social position counted for nothing. Angling skill and persistence were all that mattered.

The Tokaanu Hotel was a different matter. It catered to all travellers, not merely anglers, and provided a higher standard of accommodation than the riverside camps. The hotel itself was part of a 3-acre site, where a variety of fruit and other trees planted in the grounds sloping down to the Tongariro River made for pleasant surroundings. By 1905 the hotel had expanded to include 21 rooms for tourists.[4]

In terms of the quality and quantity of trout, the peak of the Taupō fishery, and the Tongariro River in particular, is said to be the 1923–24 season. But the first few years after the introduction of rainbow trout were almost equally as spectacular. In the first decade of the twentieth century, the lure of giant trout enticed a small group of overseas anglers to make the long journey by steamship to New Zealand.

Most of these anglers, many of them from the United Kingdom, would make annual pilgrimages. Because they had to travel across half the world to get their New Zealand fishing, they naturally stayed for lengthy periods. It was the practice of English and Scottish fishermen to arrange their summer camps at the delta for two months or more; usually sometime in the period December to April. Boats and local attendants recruited from Tokaanu were also necessary, because many trout were taken at the delta itself.[5]

During the 1890s, they had slaughtered the brown trout in the lower river in their hundreds. According to Joseph Buckingham, once the rainbows were established, Arthur Dinham Shilson took trout by the ton in the 1910–11 season, with average weight of trout caught by the party being 10lb.[6]

During the very earliest years of the Tongariro trout fishery, a small group of anglers had wonderful fishing to themselves, but had to endure many hardships as well. Malcolm Ross, who helped liberate trout in the river's tributary streams, sketched the reality of riverbank life some 30 years later. Camp facilities were basic:

> Our cook house in this camp was a rude affair, consisting of a canvas shelter, a table, and an earthen chimney. It was also our dining room, and when rain came the canvas leaked so it was advisable to wear one's mackintosh. Once the shelter blew down, and we had difficulty in cooking a hot meal in the rainstorm that accompanied the gale.[7]

There was no piped water, so bathing was done in the river — close at hand but ice cold — unless the angler made a trek to Tokaanu where a hot spring was available. Anglers could put up with most of the rudimentary living conditions, except for the mosquitoes. The lower camp in those days was close to a swamp, and at night millions of these insects swarmed to attack the anglers who were without mosquito nets. Kerosene on hands and face proved to be an effective repellent but gave only temporary relief. As for the fishing, Malcolm Ross noted there was more than enough:

> We made some great hauls of fishes — browns and rainbows — which we either buried in the sands or left on the shore for the wild pigs to eat after we had satisfied our own very moderate wants.
>
> . . .
>
> Those were the happy days when we had miles of the fishing to ourselves, and the reel of the rich tourist was not heard in the land.[8]

Admiral Sir William Kennedy RN (Rtd) was one visitor to give his impressions of the delta fishing. This 72-year-old Victorian-era naval relic had fought in the Crimea as a young officer, finding time to shoot tiger in India and pursue Swedish salmon before realising a lifelong ambition to fish for trout in New Zealand. In 1911 he made his first, and only, visit — a state of affairs which allowed him to write frankly about the country and its people. In so doing he fired something of a verbal broadside.

Writing in *The Dominion*, he reported on his one-month stay at Robert Jones's Delta Camp, where he and his nephew took many rainbows, averaging 8½lb. The trout they caught were easily comparable to salmon and just as good to eat, but the admiral condemned the delta fishing practice of letting a long line out and waiting for a take: 'to call it fly fishing is absurd, it is a prostitution of the art and more like worm fishing.'[9] He also found that the fishing, although exciting, could also be monotonous as the fish had definite times of day when they were active, while at other times an onshore wind would blow in making fishing at the delta impossible.

Kennedy found camp life pleasant enough. Jones made sure the guests were comfortable, and there was a good cook on hand. Each morning the host would arrive with milk and provisions, taking away fish the anglers had caught for distribution to those local residents who wanted them. Four rods would take some 60–100lb of trout each morning — far too many for the camp inhabitants to eat themselves. After a time, monotony set in, with fish activity dying away in the early afternoon and generally restarting around 5pm. If a breeze got up, it would bring breakers to the delta fishing spot and the fish would move out to deeper water.

The admiral paid a fee of £1 for his fishing licence, the same as New Zealand resident anglers. He found some opposition to foreign fishers coming to the Taupō region — 'damned Englishmen' — to spoil the sport of the locals, and agreed with them that a higher licence fee for overseas visitors should be charged. The fact that Kennedy had to spend some nights in the Tokaanu Hotel before a place was available in Delta Camp is some indication of the control that Robert Jones had over the Tongariro fishery.[10] John Asher, who took over the fishing lodge business after Jones died in 1924, later claimed that guests at the camp would have access to private fishing water on the Tongariro River.[11]

Charles H. M. Thring and his wife, Anne, stayed at Hut Camp in 1912. Anne gave a sketch of camp life, including an encounter with another angler there, Major-General Sir Arthur Dorward (retired), formerly of the Royal Engineers:

> There is one charming person here, Major Gen. Sir A. D. He is a little over sixty, a keen fisherman, and seems to delight in giving M. points in fishing. They go out fishing together every day.
>
> The General has more tackle than any one man ever possessed before — such beautiful flies! He has the greatest sense of humour and is the life of our little party when we gather here in the dining-sitting-room after dinner o' nights. There is a big open fire and a big table.
>
> The ti tree is tall here, and gives the camp shelter. We are having lovely weather — hot in the middle of the day and quite cold at night, with heavy dews. M. and I have two tents, as it costs no more, and he has so much tackle and guns and things!
>
> . . .
>
> Captain Ryan, over in Taupo, is a queer study. Think of that bluff, rough man having been at Julian's in Paris, and doing beautiful things in oils! People are so interesting! There is our dear old General with his title and his D.S.O. spending weeks here trying new flies, putting up with all the little 'outs' of camp-life far better than most, and as simple and happy as a child! M. is enjoying it immensely — not such big fish, but more skill required to get them — quite a different kind of fishing — that's why he wanted to come over here. It is all different. Here we have wooden floors to all tents and a wooden shack 18 by 12 feet to eat and sit in.
>
> It is quite understood that each is to do exactly as he pleases — no one says, 'Why don't you do this?' and even M. ceased to urge a big breakfast upon me! Cook sets tea and bread-and-butter on the boards outside my tent at 7a.m. and then I'm left alone until after M. has his breakfast, when he and the dog escort me to the bath.[12]

Jones's fishing (and promotional) skills

When it came to fishing, Robert Jones was one of the best. Spinning tackle could lawfully be used in Taupō rivers at the time, and on many occasions he demonstrated his deadly skill with a carefully placed Devon minnow. Like all anglers who live near a river and study it

closely, he had the knack of being able to catch trout when all around him were fishless. Sometimes, when supplies at the camp ran short, the fish he banked were destined for the guests' table.

To advertise the attractions of Tokaanu trout, Jones indulged in advertorial press notices listing the large numbers of fish his guests caught and their impressive sizes. There were frequent newspaper reports from 'a correspondent' in Tokaanu boasting of some impressive catches:

> Mr R Jones, of Tokaanu, who is evidentially an enthusiastic angler, writes informing us that during last season (covering the months of January, February, March and April) he took 76 fish from the Tokaanu River, . . . The largest taken scaled 25 ½ lb, while the average weight of the 76 trout was rather over 12lb.[13]
>
> Fishing with fly and minnow, the following catches of brown trout have been made here: — Mr McCulloch and two friends, 2 days, 48 fish, averaging 12lb; Mr Clark, Dunedin, 3 hours, 4 fish, averaging 17lb; . . . Mr R Jones, Tokaanu, 3 hours, 14 fish, averaging 17½lb . . . Shoals of brown trout are to be seen daily in Lake Taupo and the Tongariro river.[14]

Early fishing access

Travel and accommodation were not the only expenses for visiting anglers. Where they could, Māori landowners also charged fees for the right to cross their land. Controversy surfaced in November 1909 when the Tokaanu police constable reported that local Māori had decided to impose an access charge of 2 shillings and 6 pence per week for anyone wishing to fish the Tongariro River. Any angler refusing to pay would be stopped from fishing. Part of the problem lay in the fact that tourists were told that payment of a £1 whole-season fishing licence fee allowed access to all rivers. The government view was that the ban on charging for fishing rights in the Fisheries Act 1908 applied to Māori land as well as to European land. Māori lands were not private waters, so the charges were unlawful.

Jones struck a deal for exclusive fishing rights over the lower Tongariro pools from iwi chief Hoani Te Heu Heu and others who owned the riverside land. After World War I, as the fame of the Taupō fishery grew, these landowners had earned much income by charging anglers for the right to cross their properties to get to the river. (The same practice was carried out further up the eastern side of the lake by owners of land bordering the Waitahanui River.) It was necessary to charge for the right to cross land (in effect legitimising a trespass) to avoid breaching section 89 of the Fisheries Act 1908, which made it unlawful to sell or let the right to fish in any waters.[15]

In taking over the fishing rights, Jones made a wise move. For several years before he died in 1924, government proclamations had been issued prohibiting the sale of land adjacent to Lake Taupō tributaries while negotiations between the government and Ngāti Tūwharetoa were in preparation.[16] The New Zealand Acclimatisation Society also pressed for the government to acquire land around the lake to enable fishing access.[17]

Until the law was changed, though, Robert Jones had what amounted to a private fishery in the Tongariro's lower reaches, which were the only parts of the river practically accessible to anglers. Writing in his guidebook, the Auckland angler F. Carr Rollett cautioned that there were only about 6 miles of river in which to wet a line:

> [The Tongariro] runs for over thirty miles through beautiful scenery, but its banks are covered with dense scrub and the angler simply cannot get down to its beaches.[18]

In those early days, almost all available trout fishing in the Tongariro was downstream from the present location of the Tūrangi–Taupō road bridge. The waters above that point, such as the Island and Major Jones pools, were on land owned by the Downs family. Some anglers, like Zane Grey, were willing to pay a fee to access the upper pools by relying on these landowners for access. Even so, local knowledge was essential to find the paths which led through almost impenetrable vegetation down to the water's edge.

Even if an angler staying at Delta Camp had a pool to him- (or her-) self, the day could be ruined by people who trolled for trout from boats in the lower reaches. There was nothing in the regulations of the day to prevent fishing in that way.[19] Fishing in Taupō rivers from a boat was eventually prohibited, but the problem of interference by boat traffic on the lower river would continue into the 1930s.

Post-World War I years

Jones's operation continued to expand in the years after World War I. By 1920 it included the Tokaanu Hotel, fishing camps, a general store (stocked with trout fishing tackle from England's leading makers), a billiard saloon, a bakery and a transport service. In September 1923, when the general store burned down, newspapers noted that it was one of 16 buildings Robert Jones owned.[20]

Robert Jones became one of the most influential businessmen in the Tokaanu area — if anything happened, he would either have a hand in it or benefit from it. For example, in 1921 he was an umpire in the only international cricket match ever played in Tokaanu, when a team captained by Admiral of the Fleet Lord Jellicoe played a local selection on 15 February.

After Robert Jones died in 1924, his wife's brother, John Asher, was appointed executor of his will, and carried on the fishing lodges and hotel operations as managing trustee. Asher's involvement eventually led to financial complications and a messy court case, which are described in Chapter 14.

In the years leading up to World War II, Delta Camp continued to attract anglers, many of them from England, Scotland, Australia and the United States. Trout weight in 1937 and 1938 was a very respectable average of 5–6lb, with the occasional 10lb fish being taken.

James Wilson took over Delta Camp at the start of the 1938–39 season. He ran it successfully, but the venture came to an end in 1941 when control gates were installed on the Waikato River outlet at Taupō, raising the lake level by 3 feet. This meant Delta Camp was almost completely submerged, with water entering the camp huts. Jones Island, on which the camp was originally located, was swallowed by the lake and now forms part of Stump Bay.[21]

CHAPTER 6:

Decline and recovery

A consignment of about 60 Taupo trout, totalling 224 pounds in weight, has been received in Auckland, and readily disposed of through the Municipal Fish Market. Further consignments are expected from time to time.

— *THE NEW ZEALAND HERALD*, 25 FEBRUARY 1920

IN THE EARLY YEARS of the twentieth century, it was common for anglers at Rotorua and Taupō — the so-called 'thermal fisheries' — to take trout in such massive numbers that catches were measured by the ton over a season. Reports from Lake Rotorua recorded 1,167 trout as being caught for the week ending 13 November 1905, a total of 4,785lb. One year later, trout from 5lb to 9lb in weight were being taken from Lake Taupō.

In 1905, Australian angler Charles Potts and his wife, Blanche, landed some 2½ tons of trout over two months without exerting themselves, while other anglers appeared to catch more. By the end of that decade, the average weight of Taupō trout was even greater. Even photographs taken to record the size of the best specimens prompted disbelief.[1]

Trout condition declines

However, towards the end of the first decade signs were appearing that these days of free-for-all massive slaughter would not last. Anglers began to notice an increase in the number of poorly conditioned trout they caught, and that these fish contained a small, red, threadlike worm in their flesh. The average weight of fish taken also declined perceptibly. In 1913 Captain Thomas Ryan noted the average weight of Lake Taupō trout taken in the 1912–13 season was 6¾lb, down from 9lb in 1909.[2]

At this time, the administrative structure for maintaining the world's most valuable recreational fishery was a clumsy three-way affair. In 1906 the government had handed control of the Rotorua and Taupō fisheries to the Department of Tourist and Health Resorts, which administered them in conjunction with the Marine Department and the Department of Internal Affairs. This change caused consternation and outrage at the Auckland Acclimatisation Society, which stood to lose the revenue it earned from the sale of fishing and shooting licences. The society had played a major role in establishing the trout fishery, with little financial assistance from the government. The society petitioned Parliament seeking to reverse the decision, but the government, with its eyes fixed on the increasing licence revenue accruing annually, did not budge.[3]

The Rotorua lakes region attracted the bulk of tourists, thanks to its rail link with the city of Auckland, its comfortable hotels and a range of spectacular thermal attractions. And it was the Lake Rotorua fishery that was the first to show evidence of disease among the trout population, although the authorities discounted the early evidence and any detrimental trend it may have implied. By early 1908, the Department of Tourist and Health Resorts noted that constant allegations were being made that Lake Rotorua trout were declining in quality and that diseases were rife.[4] To counter these apparently unfounded comments, the department in June 1907 constructed a wire-netting weir across the Ngongotahā Stream, a major tributary of the lake, and examined the condition of the trout that were migrating upstream to spawn. Of 1,800 fish taken, 26 (1.5 per cent) were found to be diseased, with a further 270 (15 per cent) in poor condition.

That survey appeared to be sufficiently conclusive to counter the views of the pessimists. Nevertheless, the continued good health of the fishery remained a live issue. A year later officials admitted that Lake Rotorua was overstocked, but continued to play down anglers' fears that poorly conditioned fish were present in large numbers.[5]

The department was also aware of the steady decline in trout food items, such as the freshwater crayfish (kōura) and small fish in the lake. A search for possible replacement food sources began. The first attempt was made with the American cisco (lake herring), but the parcel of ova did not survive the journey. Importations of freshwater shrimp and New Zealand whitebait then followed, both obtained from the Waikato River. Once established in the Rotorua hatchery, these species looked more promising. No one knew that the shrimps would never be established in the inland lakes because their lifecycle requires them to spend time in the ocean.[6]

Overseas visitors were not slow to voice their concerns at the presence of poorly conditioned trout in the Taupō region. Controversy erupted in May 1912 when English angler Major-General Sir Arthur Dorward told press reporters[7] that every Taupō trout he caught, even those in perfect condition, contained a worm parasite, while the average weight of his catches from the Tongariro and at Tokaanu was less than in previous years.

Dorward, of the Royal Engineers, was an old China hand who had visited the Taupō district for several years. By today's standards he had nothing to complain about. After fishing for three months (February–April) in 1912, he found the average weight of trout from the river was 8lb; some 12oz less than the average weight of fish taken two years before. Other anglers, he said, told the same story; poorly conditioned trout and the invariable presence of the parasitic worm.

The general's comments were not simply about poor fishing, but directly criticised the tourist authorities for their neglect of the fishery and the lack of expert management techniques.[8]

By 1912, fisheries scientists suspected that the parasite was a nematode worm which passed from the shags (cormorants) to the cockabullies and other small fish species in Lake Taupō.[9] The worm continues to be present in some Taupō trout today, but in very small numbers. In 1995, fisheries officers described its lifecycle and its effects:

> [The] nematode (*Eustrongylides* sp.) . . . develops within the stomach of the shag, releasing fertile ova which are passed out in the shag's excreta. If dropped into the water the ova develop into larvae which are eaten by small fish such as bullies. These in turn are eaten by shags and the cycle starts again. Occasionally, however, the bullies are eaten instead by trout, which in turn become infected. The worm causes a large watery filled cyst to develop within the body wall of the trout

which, when approximately the size of a golf ball, ruptures through the skin ultimately causing the death of the trout.[10]

Not every tourist was pessimistic. Other overseas anglers continued to praise the Lake Taupō fishery as the best in the world, measuring their catches by the ton, although even they had to admit that the parasitic worm was present on occasion. Mr J. Turner-Turner, another Englishman, also found the Taupō fishing very hard in the first few months of 1912, but this was largely due to a lengthy dry spell. Once the rains came the fishing improved, although the 8lb average weight was less than the 10lb of earlier years.[11]

Colonel J. W. Moore, fishing at the same time, considered 1912 to be one of the best seasons, with only a few trout seeming to be affected by the parasite. He felt New Zealand still offered magnificent sport.[12]

The Ayson report

The Department of Tourist and Health Resorts continued to be criticised for its failure to foresee and deal with the problem. So two years after the initial survey in 1908, which had indicated there was no cause for concern, the department asked Lake Falconer Ayson, Chief Inspector of Fisheries for the Marine Department, to assess the situation. By that stage, press reports of diseased trout were turning up in significant numbers. Some people felt that Lake Rotorua held too many trout, limiting their growth, and Lake Taupō would go the same way.[13]

Ayson warned in a 1910 report that Taupō was overstocked with trout, and the food supply had to be safeguarded:

> To do this large quantities of trout should be taken out before any serious depletion of the natural feed takes place. If this is not done promptly there is little doubt but that the Taupo fishery will quickly deteriorate. To reduce the number of trout and utilize them to the best advantage I would recommend netting portions of the lake, and in such a way *as not to interfere with angling*; and marketing the fish. . . .
>
> The netting should be done thoroughly and the trout so reduced in number that there would be no question of the further depletion of food supply.[14]

The state of the Taupō fishery reached the highest level of government when Cabinet discussed the matter on 22 May 1912. Experts advised that the decline in fish quality was likely temporary, possibly caused by bad summer weather limiting the food supply for the trout. Government ministers decided on a two-point plan: poorly conditioned trout should be netted (the fishing season was about to end), and the extermination of shags should be encouraged through a bounty being paid for each shag killed.[15]

In mid-August 1912, Ayson visited Taupō again in order to make a report to the Marine Department. Although the evidence from anglers was only anecdotal, the people Ayson interviewed were all long-time residents of Taupō and Tokaanu who had been present since trout were first introduced to the district. Most agreed that the trout had suffered a noticeable decline in size and condition over the previous three years, small food fish such as the kōkopu had largely disappeared, and trout numbers had increased. While the point as to an increase in numbers could only have been a subjective impression, Ayson's broad calculations appeared to confirm it. With the spawning streams feeding Lake Taupō exceptionally well-suited to the propagation of trout, and an absence of natural predators such as eels, it was inevitable that the numbers of fish killed each season by anglers (estimated at 100 tons or 56,000 trout) could have little effect on the annual population increase.[16] The trout he inspected in the spawning streams had an average weight of 5–6lb.

Ayson's conclusions in 1912 were essentially the same as before: the Rotorua and Taupō fisheries should be placed under the management of fisheries experts at the Marine Department, trout numbers should be thinned out, diseased fish destroyed and shags eliminated, while the introduction of a suitable small fish or other food item would also assist. The entire project had to be carried out in a careful, methodical manner.

Ayson also commented on the increasing effects of economic forces. Noting that Māori landowners had begun to charge anglers for the right to camp on riverbanks and to fish in certain rivers, he advised the government to acquire land on the margin of Lake Taupō and along the banks of the best fishing rivers. But the quality of the fishery was the immediate problem, and the issue of angler access was to simmer for another 14 years before the government dealt with it.

While the Department of Tourist and Health Resorts did take action, its initial focus was oddly misplaced. Instead of adopting

Ayson's two-pronged solution of netting the trout and boosting their food supply, officials accepted the opinions of government scientists that the worm parasite was the cause of the decline in trout condition. Even the Conservator of Fish and Game, Frederick Moorhouse, thought that nothing could be done until the nature of the parasite had been ascertained, Ayson's conclusions about culling and food supply notwithstanding.[17] The department therefore launched an attack on a symptom and not the underlying cause. In reality, the lack of condition on the part of the trout stemmed from overpopulation and the consequent reduction in available food. The fish had simply exhausted their larder.

With the outbreak of World War I, New Zealand scientists had to put off research that would solve the nematode worm problem. It was not until 1917 that they ascertained that the trout was the intermediate host for the parasite, while the shag was the ultimate host in which it grew to maturity. The worm never left the trout's body, relying instead on predation by the shag to transfer to the final stage of development. Trout infested with the worm experienced a significant decline in health, making them easy prey for fish-eating birds.[18]

Professor Prince's visit

A government faced with an intractable problem will often call in an overseas expert for advice. This strategy has political advantages; if the advice is sound the government takes the credit, but if it fails then the expert gets the blame. (In either case, the expert's financial reward is invariably substantial.)

In June 1913, after receiving L. F. Ayson's report, the government invited Professor Edward E. Prince, the Canadian Fisheries Commissioner, to review the country's coastal and freshwater fisheries and advise on how the resource could best be managed and developed. As cynics have observed, consultants are an expensive way of finding out what is already known, and Prince's visit seemed to follow that pattern.

The professor arrived in New Zealand in early April 1914 to a ministerial welcome. He spent three weeks on a steamer inspecting coastal fisheries, before travelling to Taupō and Rotorua to study the state of the trout fisheries there.

Four months later, Prince set out his impressions in a wide-ranging 'preliminary report', which covered everything from minnows to marlin and whitebait to whales. In the section discussing trout and salmon, he endorsed the work of New Zealand fisheries scientists on the worm parasite (which he felt had run its course), and recommended further netting to reduce the trout population in the lakes. He also advised strongly that one government department should have responsibility for all fisheries administration. These points really added nothing to the data that Ayson and his colleagues had already provided. Prince then collected a tidy £1,000 fee for his work and departed, promising to provide a final, more comprehensive, report. Four years later the government was still waiting for it.[19]

On 1 June 1913, the Department of Internal Affairs took over management of the Rotorua and Taupō trout fisheries, while the Department of Tourist and Health Resorts continued to act as the acclimatisation body, issuing licences to anglers. Netting in Lake Taupō did not start until 1 July 1913. By delaying the netting programme and other measures recommended by L. F. Ayson, the authorities allowed an unsatisfactory situation to continue for far longer than was necessary; a point which Professor E. E. Prince made in his 1914 report.[20]

Culling diseased and inferior trout was just one part of a wider range of measures used to improve the thermal fisheries. The rest of the plan involved culling and selling healthy fish (fresh or smoked) to the public, researching causes of the worm parasite, eliminating shags, and introducing new trout food items.

Meantime, trout fishers had become impatient with the apparent lack of action by the authorities. The Rotorua Rod and Gun Club circulated a petition calling for control of the 7,000 square miles of the thermal fisheries district to be returned to an acclimatisation society operated by local volunteers instead of paid government officials. Government control, the petition alleged, had been a failure because there was 'no definite practical scientific policy'. Amateur enthusiasts would succeed where the plodding efforts of the bureaucrats had failed.

The club officials presented the petition to Parliament on 18 July 1913. A parliamentary committee heard evidence from interested parties later that month. Although the matter occupied several days, the Minister of Internal Affairs, the Hon. Francis Bell, had already indicated that the government was not going to relinquish a prized dominion sporting asset. On 24 July 1913, the acclimatisation societies' lawyer Charles Morison KC (an angler himself, from Wellington) presented a remit advocating that control of the Rotorua fisheries should be passed from the Department of Tourist and Health Resorts to the societies. The minister brusquely dismissed it, arguing the large sums involved were incompatible with local control.[21] Fisheries management on a large scale was not a matter for amateurs.

The netting programme

The netting of trout at Lake Taupō started in 1913, using traps and drag-nets rather than erecting barriers across the mouths of the rivers and streams running into the lake. The practice of the Department of Internal Affairs was to target ill-conditioned trout for destruction in the closed season (1 June to 30 October). The marketing of fresh and cured trout to the public occurred only during the period November to March.[22]

Not all fish caught for sale were fit for human consumption, of course, and statistics kept indicate the numbers removed from the market catch during the open season. The cured trout found a ready market in Auckland and other centres.

The following table shows the numbers of trout taken from Lake Taupō for the seasons between 1913–14 and 1919–20.[23] The first figure listed for each year represents trout taken and sold at market (either fresh or smoked). The second figure is the additional number culled from the market catch for lack of condition. For the years from 1914 to 1917 figures from the pre-season culls are included.

	Trout sold at market	Poorly conditioned fish culled
1913–14	6,243	2,830
1914–15	11,574	12,779*
1915–16	16,137	15,674*
1916–17	8,743	8,104*
1917–18	17,947	5,294*
1918–19	10,060	2,385
1919–20	11,493	780

*Includes poorly conditioned trout culled in the pre-season period.

The scale of the netting activities can be indicated by weight. During the first three years of netting by the Department of Internal Affairs at

The government netting programme undertaken in Lake Taupō sought to balance fish numbers with the available food supply.
Auckland Libraries Heritage Collections AWNS-19140212-47-1

Taupō (1914–1916), 68 tons of poorly conditioned trout were removed, while trout sent to market weighed 55 tons for that period.

Netting Taupō trout on a large scale would seem to be an easy thing to do, but results were mixed. The department's launch could not put out in bad weather, and rivers were often in flood, restricting operations. There was also a shortage of manpower because of the war effort.

At first, there was little visible difference in the quality of anglers' catches. Visiting English angler A. Millington Naylor had observed the decline in trout weight for six years, beginning in 1908 when the average weight of catches had been 9¾lb. By April 1914, Naylor reported the average was down to 4lb, with many fish being in poor condition, while angler numbers were reportedly less than half those of the previous year.[24] The following year he found the trout were about the same weight but of better quality.[25]

By April 1917, the netting of trout over in Lake Rotorua was showing its effects. Officials reported that the fish there were markedly improved in condition, and the incidence of the worm parasite was decreasing. The Auckland City Council established a fish market to which cured Rotorua trout were sent by rail, some being used as food on the train dining cars. Auckland boarding houses also took trout on a daily basis.[26]

Netting in Lake Taupō, however, was a more difficult assignment. It had taken three years to reduce the trout numbers in Lake Rotorua sufficiently, and Rotorua was a smaller, shallower water body where netting could carry on unimpeded by the weather. Lake Taupō was a different proposition; much larger, and exposed to the prevailing southerly winds. The two fisheries launches could not operate when the lake was rough, and there were few places suitable for drag-netting because most river mouths had obstructions, such as rocks and sunken trees.

No fish were culled in the seasons from 1920–21 to 1927–28 because of high labour costs and difficulties in transporting the fish to the railhead, but the netting was nevertheless clearly producing results. In his report for the 1921–22 season, the Conservator of Wildlife stated that fish were being taken from Lake Taupō up to 15lb, and the average weight of the trout was 7lb. The liberated food fish (not identified) had taken hold and were evident in their millions, so all the indications were that netting for two more seasons would return the fishery to its former high standard.[27]

The process resumed in the 1928–29 season using traps in the spawning streams. Out of the 9,000–10,000 trout handled at the traps, an estimated 4 per cent (up to 400) were destroyed for reasons of age, poor condition or deformity. The numbers removed were nothing like those of the previous decade.

Trout condition improves

As the seasons passed, the netting had the desired effect. The trout population became better balanced with its food supply, and poorly conditioned fish were fewer. While staying at the Delta fishing camp in 1919, New Zealand angler Bartley E. Durrant took more than 224lb of trout in one day, fishing from a boat. His largest rainbow trout weighed 9lb, and he caught many of 6lb.[28]

By 1920, the recovery of the fishery was well underway. 'No fool could fail to catch fish in Lake Taupo,' said Sir Walter Essex after an afternoon's trolling provided trout ranging from 3lb to 5lb.[29] Another English angler who kept records of his catches in the 1922 season noted that the trout he caught had an average weight of 7lb, an appreciable increase on the 3½lb average of fish taken in 1919.[30]

The fishery managers took other steps to retrieve the glory days of the past. The programme of exterminating shags continued, with 645 of the birds being destroyed in the 1920–21 season.[31] It was also thought that the thermal fisheries needed fresh blood, so a consignment of 250,000 rainbow trout ova was obtained from a hatchery at Lake Hāwea. The fry subsequently hatched were liberated in Lake Taupō and Lake Rotorua, in the belief that this strategy would improve the strain of trout available to anglers.[32]

By late 1922, the tourist authorities could confidently report that: 'The trout fishing during April and May of 1922 was of the usual high standard of the past two seasons. The trout worm or parasite which affected the rainbow so badly from five to fifteen years ago has become almost non-existent.'[33]

Wellington angler Frank Dyer found the fish hugely improved in condition. Fishing on the Tongariro in May 1923, he had bagged trout with an average weight of 8lb.[34]

The Taupō trout fishery, and the Tongariro River in particular, were on the threshold of worldwide fame. In the 1923–24 season, with the worm infestation successfully kept in check, the average weight of Taupō trout reached its all-time high, with reported catches (described by one newspaper as 'salmon trout')[35] having an average weight of 11lb and 12½lb.[36]

The great Tongariro fish of the 1923–24 season fully merited salmon tackle. The Australian angler R. Johnstone quickly learnt this fact when he came to Tokaanu that year. In giving a brief summary of his impressions, he also set out the basic requirements for visitors:

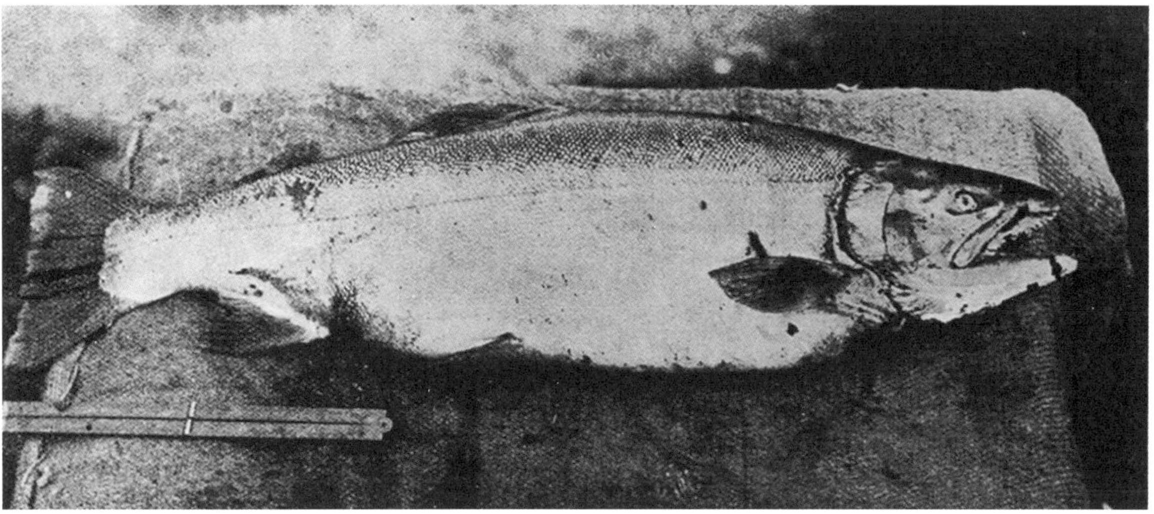

An example of the massive trout caught in the 1923 season. This rainbow specimen was caught in Lake Taupō's Acacia Bay.

The Manawatu Standard, 31 March 1923, p 4. Auckland Libraries Heritage Collections AWNS-19230412-49-6

> . . . one needs a rod of at least 13 ft, a good free running reel containing at least 120yds of good strong silk line, and the flies should be the very largest salmon flies obtainable.[37]

Johnstone found his 11-foot Hardy fly rod proved too light for the Taupō trout, which would take the fly and disappear downstream using the river's strong current. Switching to a rod of 18 feet made all the difference. Johnstone described the Tongariro fishing methods thus:

> Wading well out to the centre of the river, or as far as you can safely go, you throw your fly well across to the opposite bank. Of course, as you are using a big rod, you will use both hands for casting. The strong current will take your fly down stream, and you will keep paying out the line very slowly, working the tip of the rod up and down all the while. If you are in luck you will feel a heavy pull, and your rod will almost straighten itself. Strike hard, see that your line is free of the reel handle, and make the best of your way to shallow water, where you can play your fish in comfort, and, if necessary, follow him. I have never experienced anything like the thrill you get when one of these big rainbows take hold. You do not see the rise, as the fly is under water. As soon as you have hooked him he leaps right out of the water, so you have the satisfaction of seeing him, although you may lose him afterwards.[38]

Those who fished the 1923–24 season could not know it, but they were fishing in the Tongariro's finest era. 'You can't catch anything smaller than 9lb,' exulted N. S. Field, a retired judge from the hill country of Burma. 'There is nothing like that at Home or anywhere else.'[39]

Cecil Whitney and friends opened the season with 67 Tongariro trout. They made their catch with the fly rod, fishing the pools near the Taupō road bridge instead of angling from a boat at the delta. The middle reaches of the Tongariro River were now becoming more popular.

The big catches carried on into 1924. In January of that year, an angler called Cassel, fishing at the Tongariro River mouth, took 15 trout at an average weight of 12lb. Two fish weighed 17lb and one was 20½lb.[40]

The bag taken from the Tongariro River by a group of Feilding anglers in April and May 1924 is an example of the extraordinary fishing available.[41]

Angler	Days fished	Number caught	Total weight	Largest trout
J. Hurdle	3	9	92¼lb	15lb
B. Durrant	6	13	142lb	14lb
F. Stockwell	5	7	74lb	14lb
C. Dunford	4	15	141lb	14lb
N. Wright	7	20	208¾lb	17lb

Out of the total of 64 rainbows landed, 34 weighed 10lb or more.

To spread the fame of the Taupō fishery and encourage tourism, mounted specimens were put on show at the 1924 British Empire Exhibition. History shows that the high point was the 1923 season, but well into the 1930s the average weight of trout taken was heavy by modern standards, and double-figure fish were by no means uncommon.

After World War I, Taupō trout average weights reached a high of 8lb 4oz (3.74kg) in the 1923/24 season. The extreme variation in trout weight during the pre- and post-war years is shown by catches recorded in the fishing diary of Tongariro regular J.S.W. Neilson:

- 1910–11 season – 8 ¾lb
- 1913–14 season – 6 ¼lb
- 1917–18 season – 3 ½lb

Another angler, George Cotterell, took trout with an average weight of 3lb in 1917–18 but doubled that to 6lb in 1922–23. In the 1923–24 season both Neilson and Cotterell took trout at an average weight exceeding 10lb.[42] The largest rainbow trout taken in the 1920s was a 27½lb specimen caught at the mouth of the Waitahanui River.[43]

One of the largest Tongariro rainbows in this period of resurgence was a 21½lb specimen caught in 1924 by Mr Yatman, an English angler.[44] Trout fishing of superb quality continued into the next season, with Cecil Whitney and a friend taking an impressive opening day bag from the Hut Pool: in two hours they banked 26 rainbows at an average weight of 7lb. Only a few years later, Romer Grey, son of Zane Grey, would take a fine 15¼lb fish from the Dreadnought Pool.[45]

These great days were not to last. Several years on, the average weight of trout caught had dropped considerably. In April 1926, Sir Francis Colchester-Wemyss, a retired Indian Army officer and member of the Fly Fishers' Club, fished the Tongariro for two weeks with a friend. Fishing was hard because the river was low and clear, or perhaps Colchester-Wemyss used the wrong flies (he never fished a fly he had not tied himself). Even so, their bag totalled 61 trout weighing from 5lb up into double figures, with the heaviest weighing 11½lb.

Two years later, Englishman Dr Lewis Smith spent two weeks at Delta Camp. He and his wife landed 232 trout at an average weight of more than 5lb. All were caught by fly fishing, the heaviest fish weighing 10lb.[46] The figures make impressive reading by today's standards, but the legendary monsters of the early 1920s were absent.

An Australian fly fisher, Dr A. Allan, visiting the Tongariro in 1927, noted the variety of nationalities and occupations among his fellow anglers: Royal Navy men, British tourists, New Zealand judges and other professionals, a small colony of Australians, and a group of New Zealand farmers. The 20 or so Americans making up Zane Grey's party, camping at Kowhai Flat, topped off the list.[47]

Scottish angler A. Mathewson took 45 trout in three consecutive days in May 1929, securing the limit each day. The estimated average weight for his catch over the five weeks of his stay at Tokaanu was 7lb per fish.[48]

Smelt introduced

In the early 1930s the authorities wanted to ensure that the Taupō trout fishery would not crash again. The Conservator of Fish and Game, Andy Kean, decided to try adding a small bait fish to Lake

Taupō. This species was the smelt (*Retropinna lacustris*), which was already well established in Lake Rotorua, a short distance to the north. The Taupō trout had largely decimated the small native fish, the kōaro, some years before, leaving the common bully and (in summer) the green beetle as the two main food items available for trout. Adding another food fish to the list would surely be beneficial, given that the smelt was thought to be a fast breeder and fed on the surface. In 1931 smelt had already been successfully transplanted into Lake Tarawera, one of the major lakes in the Rotorua district.

The date of the first smelt introduction to Lake Taupō occurred in the mid-1930s, either in 1934 or 1935. The exact date is unclear, but the first experimental transfer was most likely in 1934. Kean was later to state that he made the first liberation on 12 May 1934, in the large bay at the head of the lake where the Waikato River begins its journey to the sea.[49] Fisheries field officer Pat Burstall records the date as 12 July 1934, but in the same location.[50]

The trial runs Kean made that year with small batches showed the species would survive the journey, provided adult fish were used. More consignments were accordingly shipped in the following years. The smelt seem to have spread throughout the lake by first concentrating their numbers along the western shores; possibly because that location gave better protection from the prevailing wind. The smelt were thought to be breeding successfully by 1939, although Kean's reports to his superiors at the Department of Internal Affairs show a measure of cautious optimism. In late November 1939 he pointed out some promising signs — the smelt were plentiful and breeding, and appeared to be larger than those in the Rotorua lakes — but he would not confirm the species had successfully been established. By March 1940, Kean was able to report that he had observed smelt visible in great numbers on the eastern side of the lake, while reports from Western Bay indicated the species was equally prolific there, as well as at the Tongariro River delta. Not only were the small fish well established, but the trout were avidly feeding on them near the surface.

By that stage the total number of smelt introduced was approximately 278,000. It was obvious no more shipments of smelt from Rotorua were necessary. Anglers were not slow to adapt to the new food item, and began fishing with smelt flies modelled on the patterns used in the Rotorua lakes.

Along with smelt, the Wildlife Department decided to try another food source in the form of freshwater crayfish (kōura). More than 4,000 were added to Taupō waters in 1937–38.

The final seal of approval came from one of the Tongariro River's old hands J. S. W. Neilson. Writing in mid-1940, he congratulated Kean, recounting the fine condition of the trout caught from January to the end of the season. A noticeable feature was that inferior trout were very few in number in his catches. Neilson wrote:

> I am convinced that this great improvement is due to your establishment of the smelt in Lake Taupo. I saw countless shoals of smelts at the Tokaanu wharf and in some cases along the beaches, and I could see trout feeding and chasing them a long way out on the lake.
>
> . . . I have fished Taupo and especially the Tongariro River since 1910–11, and I can say, except for the numbers of rods now on the popular pools, that I have had as good sport as ever I had, . . . on occasions this season I have had for the day an average of 6¼lbs. in seven fish . . .[51]

Kean echoed Neilson's comments some months later in September 1941 when he confirmed the project had been a 100 per cent success, backed up by evidence of an increase in trout weight. Pat Burstall, writing years later and with much experience of the Taupō fishery, was certain that smelt were the salvation of the Taupō rainbows. However, while smelt supplied an important food source for the trout, this came with one disadvantage for anglers. Before smelt appeared, Taupō trout had been marginal feeders available to anglers around the shores of the lake all year round. Smelt are pelagic, living mostly in Taupō's open water. Their preference for that zone changed the distribution of trout around the lake, with fewer trout available to shore-based anglers.

In its 1937 annual report, the Department of Internal Affairs noted that only 11 fish were culled from hatchery operations on Tongariro tributaries;[52] a sure indicator of the quality of the fishery.

CHAPTER 7:

Zane Grey in New Zealand

What is New Zealand worth to Zane Grey, and what is he worth to us?

— ***NEW ZEALAND TRUTH*, 5 MAY 1927**

IN JANUARY 1926, the SS *Makura* docked at the waterfront in the city of Wellington. Among the passengers was a middle-aged North American man with a determined gaze and silver hair parted neatly in the centre. The passenger was Zane Grey, writer and outdoorsman, who had come to New Zealand to sample the sport fishing opportunities. While his main focus was big game fishing (sharks and billfish), his experiences with the Tongariro trout were to leave an indelible impression upon him.

The novelist/angler was born Pearl Zane Gray (later changed to Grey) on 31 January 1872. The son of an Ohio dentist, he learnt to fish in his boyhood, largely through his own efforts because his parents disapproved of outdoor pursuits. After attending the University of Pennsylvania he worked as a dentist in New York, but the call of the wild was too strong and he began writing for *Field & Stream* and other outdoor magazines. Although self-taught as a writer, he had enormous determination to succeed, and ultimately found that the American public was hungry for fiction in the form of the adventure western.

His first contribution to a sporting magazine in 1903 was a fairly tame one about camping out for which Grey had to supply his own illustrations.[1] Once established as a successful author, he could afford to take photographers and film cameramen on his fishing expeditions to record his feats. There was even a camp cook, George Takahashi, who became a regular member of the party.

By 1926 Grey's reputation was at its zenith. Almost 54 years old, the author of such novels as *The Thundering Herd* and *Riders of the Purple Sage* had perfected the art of mixing action, light romance and frontier justice in the right blend for an audience that was highly receptive to the Western genre. While critics complained that his works were shallow and formulaic, no one could argue with the financial returns Grey garnered, particularly once his books were turned into successful motion pictures, ensuring even greater exposure. In 1921 his income had been almost US$128,500, a substantial amount for that time. In 1924 he more than doubled that amount to nearly US$292,500.

However, the public persona Grey cultivated as an adventurer hid a complex man, egotistical and strong-willed. He was convinced the key to his success as a writer lay in the rejuvenating presence of attractive young women, together with a degree of freedom from his family in order to enjoy the outdoors. A teetotaller from an early age, he also placed great stress on personal loyalty from his associates — a rather hypocritical position given his penchant, once married, for keeping a mistress and being a serial philanderer. Periodically, he suffered from severe depression when emotional setbacks occurred after one or other of his younger female companions took other partners.

At the time of his 1926 visit he was involved in an emotional correspondence with Nola Luxford, a New Zealand actress then residing in the United States. This was one of the few encounters which Grey did not initiate, but he was captivated nonetheless. The affair, if it could be called that, fizzled out when he returned to the United States, largely because he and Nola Luxford had different objectives. Far from engaging in a brief and diverting emotional dalliance, she was trying to keep her struggling film career going.[2]

Zane Grey invited

Fishing journals had long extolled the exceptional qualities of New Zealand trout fishing. Zane Grey first learned of it in 1900, and, as

Charles Alma Baker, the cattle rancher and rubber plantation owner who helped to bring Zane Grey to New Zealand.
Auckland Libraries Heritage Collections AWNS-19260128-39-5

with any new, untested fishing ground, longed to try it for himself. The 1926 visit came after Charles Alma Baker, a New Zealand-born businessman, visited Grey at the Tuna Club on Santa Catalina Island, California, in 1925 and passed on an official invitation from the New Zealand government.[3]

Baker was a retired rubber plantation owner from the Malay Straits Settlement, a Crown colony made up of trading centres of which Singapore was one. He had pioneered the planting of rubber in Malaya in the late nineteenth century. When an investment boom occurred in 1910, the returns from Baker's plantations ballooned and he became immensely wealthy. By the 1920s, he already had a reputation as a big fish angler, and had fished with Grey regularly in Honolulu and the waters off California.[4]

The New Zealand government, wishing to boost its tourist revenues, consulted Baker on how best to publicise the country's sport fishing overseas. The 'Rubber King', as the newspapers dubbed him, nominated Grey as a possible solution, and in effect acted as an envoy.[5]

The New Zealand authorities wanted Grey to publicise the glories of New Zealand's angling, in return for which they agreed to give him transport assistance in the form of free rail travel and use of the government launch on Lake Taupō. A ringing endorsement from the world's best-known fisherman, so it was thought, would surely bring in more foreign tourist dollars.

Several people (including some in the Publicity Department) later claimed credit for the 1926 visit. It would seem that the urging of Baker, backed up by Grey's own research, was the chief factor in his decision, and convinced him an expedition to New Zealand would be an enthralling angling experience, with fish of sufficient size to provide the world records he craved.[6] He and Baker had already discussed the prospects for a New Zealand fishing trip in 1924, and by August 1925 Grey had cabled Baker in London arranging to meet in New Zealand the following January.[7] The official invitation from the New Zealand government tipped the balance by promising to provide assistance with travel costs.

To ensure the great man's battles with game fish were recorded, a Publicity Department photographer was attached to the party to film the visitors in action out on the water. As a further incentive, Grey was allowed to use the government footage as part of the motion picture he would later produce.[8] The general expenses of his entourage he would meet himself.

Grey arrived in New Zealand as a well-established, highly successful author. At this time he was in the middle of a virtually non-stop series of fishing adventures. Earlier, in 1925, he had travelled to the Galapagos Islands, before making an arduous boat trip down Oregon's Rogue River, fishing for steelhead trout. He had undergone surgery shortly before the river trip, and, although it was successful, it had exhausted him. His writing had, however, enjoyed some success. His third angling book, *Tales of Fishing Virgin Seas*, had recently been published to favourable reviews. His first, *Tales of Fishes*, was published in 1919. Its colourful, well-crafted descriptions of sea fishing exploits gave notice that Grey's talents extended beyond the genre of the Western novel.

Dolly Grey

A woman married to an outdoor adventurer is forced to accept her spouse's long absences. Grey's wife, Dolly, however, had far more to put up with, because her husband, although married, behaved as if he were not. On his first New Zealand visit he took his mistress, Mildred Smith, explaining to the newspapers that she was his secretary. Dolly Grey accepted all this from her 'wandering Ulysses', as she called him,[9] and downplayed his serial romances as mere sideshows. In their letters they profess undying love for each other, while at the same time discussing how Zane Grey should deal with his mistress when she became difficult.

Despite their complicated marital relationship, the couple had a highly effective publishing partnership. Their joint efforts resembled a manufacturing firm; Zane handled production (writing) while Dolly looked after marketing and distribution. With Zane away for long periods, Dolly also managed the couple's finances, arranged book and film contracts, and attended to any other business matters. In all of these, she proved to be a loyal and extremely capable manager, alongside raising the couple's three children pretty much single-handedly. Because Grey channelled all his manuscripts through her, she became sole editor and, as a consequence, a powerful influence on the quality of his work. Before marketing the stories to publishers, she would correct the spelling and generally polish the text and ensure overall standards were maintained. Only once did he submit a manuscript to publishers without her advice and opinions. It was rejected.

Captain Laurie Mitchell

Grey's angling companion, Captain Laurie Mitchell, was born in England. After serving in World War I with the Royal Engineers, he joined anti-Bolshevik forces fighting in the Crimea.[10] The two men first met after the Great War, when Grey was fishing the waters off Nova Scotia. Their friendship was put on a firm footing in 1924 when Grey returned to Nova Scotia to fish once more for the giant bluefin tuna, a species for which Mitchell held the record until Grey overtook him.[11]

For Grey, Mitchell was the ideal angling companion. He was an experienced angler, at home at sea or on river or lake, able to join his employer on the water, yet content to stay in the great man's shadow. A photograph shows him at the riverside in canvas waders, his tall fishing hat giving him the curious appearance of a grinning gnome who has briefly forsaken his subterranean existence for a day on the water.

What Mitchell thought of Grey's womanising is unknown.

Stern countenance: Mr Zane Grey on board RMS *Makura*.
Alexander Turnbull Library, Wellington, New Zealand. /records/23068613]49-6

Presumably he turned a blind eye to it, given Grey was the one paying his salary. If his employer wanted to have a long-range *ménage à trois*, then that was, so to speak, his affair. The other drawback might have proved a greater strain for Mitchell: Grey, a lifelong teetotaller, allowed no liquor in the party.

Grey's first visit

For the Grey party, big game fishing was the immediate priority. From Wellington the group took the train to Auckland, from where after a

Captain Laurie Mitchell with a large male rainbow trout taken from the Dreadnought Pool, Tongariro River. The pumice bluff which gave the pool its name can be seen in the background.
The New Zealand Herald photo, Sir George Grey Special Collections, Auckland Libraries, 1370-U053-01

brief stop they headed north to the big game fishing centre of Russell, in the Bay of Islands. This was to be no spontaneous, Walden-inspired communion with Nature, though. There were several truckloads of camping equipment, much fishing tackle (with which Zane Grey liked to tinker) and three typewriters.

The men first spent six weeks deep sea fishing in the Bay of Islands, setting world records for marlin in the process. Grey had not encountered New Zealand deep sea fishing methods, but he had a capable local boatman in Peter Williams and soon adapted to the conditions. There was one exception: Grey began raising the hackles

Zane Grey with a fine rainbow trout from the upper Tongariro.
Sir George Grey Special Collections, Auckland Libraries, 1370-U053-03

of the other anglers in his eagerness to advertise the results of his efforts when a particularly fine fish was boated, his practice being to announce the fact to everyone within earshot.

The party then moved to the Taupō area, arriving on 31 March 1926 to find all the local hotels fully booked in anticipation of the Easter holidays. With accommodation available for one night only, and no campsite to his liking on Lake Taupō's eastern shore, Grey went to a remote bay at Waihora on Lake Taupō's western side. This location could be reached only by water, and so gave the group some seclusion for its first experience of Taupō waters. In two weeks there, Grey and Mitchell caught 70 rainbow trout at an average of 6½lb (the average weight for the season).[12] In mid-April 1926 Grey and his companions shifted base to Kowhai Flat on the Tongariro River, which they fished for the comparatively short period of six days.

Stories of large catches of trout from the Tongariro had reached the party by word of mouth and in the press, yet with the water low and gin-clear, Grey and Mitchell felt fortunate if they killed several trout in one day. Newspaper reports, richly endowed with cheerful optimism, were at odds with the reality of the river. The following example is typical:

> Fly fishing at Tokaanu during two days Messrs Whittle and Ensor secured a magnificent catch of 37 trout. This catch is easily the best bag secured in these parts this season and averaged exactly 10 ½ lb, the largest being 16 ½ lb. The condition of these fish has never been excelled, and they provided great sport.[13]

The unpalatable truth about Taupō trout fishing was revealing itself. The huge bags of trout landed each year by local anglers were taken on spinning tackle, complete with triple hooks, or on heavy fly rods, often wielded at night. The tourist publicity had refrained from mentioning those methods. As Ernest Wiffin, Grey's local contact, confirmed, New Zealand anglers were intent on catching limit bags; it was numbers above all else. Grey and Mitchell, with their 6oz rods, found it harder to raise fish in the clear waters until they changed over to the larger local flies.

Zane Grey fished almost entirely with members of his own entourage, rather than actively seeking out local anglers, although the Hawke's Bay farmer and angler A. R. (Bob) Mills, who also camped at Kowhai Flat, writes of fishing in his company.[14] Most likely, Grey felt no need to encourage contact with the locals, who saw him as a celebrity and came regularly to his camp where he could hold court.

The idyllic location of their camp and the enchanting clarity of the river could not compensate for the unfavourable fishing conditions. Taupō trout, like other salmonids, require a rising river before they will migrate upstream. A severe drought had prevailed over most of New Zealand for many weeks before the Grey party arrived, which meant the spawning runs were small, and those fish that were in the river were not taking the fly freely. Despite these handicaps, Grey explored the pools in the vicinity and caught trout almost immediately, although he found success did not come until he used a fly incorporating the feather of a local native bird, the bittern.[15]

His experience in fishing for steelhead trout on the Rogue River had prepared him well for the necessary Tongariro techniques, so it was simply a case of changing to the larger lures which the local anglers favoured. By his own account, Grey was the first United States angler to fish the Tongariro.[16] Whether this is correct may never be known, but he was certainly the first American to fish the river and then write a book about the experience.

During their stay, Māori guide Hoka Downs[17] took Grey and Mitchell to the Dreadnought Pool, a short distance above the Kowhai Flat camp. (At that time, the riverbank was privately owned by the Downs family.) This pool would enchant Grey, and he returned to it several times. With massive, commanding bluffs and thunderous rapids, its imposing location made it an entirely appropriate resting place for the large trout the men could see from the high bank. It was here that Grey took his two largest rainbow trout, each one weighing more than 11lb.[18]

He broke camp on 21 April 1926, the very day on which Prime Minister Gordon Coates was in Waihi village, at the southern end of Lake Taupō. Coates intended to negotiate an agreement between the Crown and representatives of the Ngāti Tūwharetoa iwi which would allow anglers access to Lake Taupō and its tributary streams. The agreement formed the basis of legislation enshrining the fishing rights which anglers have enjoyed in the Lake Taupō district down to the present day.[19]

Fishing technique and tackle controversy

Once Zane Grey started fishing in the Bay of Islands, he immediately noticed the differences between New Zealand and United States angling techniques and tackle. Some visitors might have felt bound to ignore these differences or to diplomatically downplay them. Not so Zane Grey; when asked by reporters, he spoke his mind. In a specially commissioned newspaper article,[20] he opened with a brief lament on the loss of northern New Zealand native forests, before condemning as unsportsmanlike the treble or 'gang' hook which the local boats commonly used in their pursuit of shark and marlin. The controversy was not helped by the American's penchant for self-promotion when fishing; an attitude which taciturn New Zealand and Australian big game fishers took for conceit.

Once the controversy over crude New Zealand fishing techniques erupted, it widened to include the locals' fly-fishing methods on the Tongariro River. At the end of his 1926 tour, Grey gave an interview to another newspaper reporter. While happy to acknowledge the superiority of Taupō trout and their magnificent fighting qualities, the beauty of the local scenery and the helpfulness of tourism officials, he also made some frank comments on Tokaanu fly fishers whose fishing methods he viewed as unnecessarily heavy. Trailing a fly on a long line and waiting for a take was not really fly fishing. To make his point, he compared his rods with those of the locals:

> We fish with five or six ounce rods with the dry fly, and even with the big fish at Taupo, and using salmon flies we had only 6½oz rods and lines and leaders (or casts as you call them) to go with such an outfit. We got 134 fish, averaging 5½lb. I caught one of 11¾lb and another of 11½lb on that light tackle, and Captain Mitchell caught several from 9lb to 11lb, all on the same light gear. Out of 12 fish hooked on the Tongariro River, I lost only one. With such light tackle, it is the finest fishing in the world . . .

Captain Laurie Mitchell was equally convinced of the virtues of light rods, which allowed the fish to fight better. New Zealand fly fishers were clumsy by comparison:

> I saw men at the mouths of the Tongariro swinging heavy double-handed 14ft rods splashing the water with thick silk lines, fishing down and across the current, and even letting the fly swim about on its own account, as though it were a line bait. Some of these fishermen got very little luck, and they could not understand why we, fishing some miles above them, landed plenty of fish. Our light tackle and smaller flies suited the clear, low water and we fished upstream.[21]

Possibly because of this outspokenness and difference of opinion, Grey was not a guest of the New Zealand government on his later visits.[22]

The angling community bristled at this broadside from their wealthy visitor. Here was a guest of the government departing from the standard script expected of foreigners about the New Zealand scenic wonderland and its equally wonderful people.

In expressing these views, Grey saw himself as an educator, providing helpful information for the benefit of local anglers.[23] But many New Zealanders objected. What Grey saw as advice, they took as unjustified criticism from an interloper, and a foreign one at that. There was a heated exchange of views in the Auckland press. Captain Mitchell did his best to explain the merits of the Grey approach in a second article, but he didn't help his case when he described the New Zealand method as 'decidedly mid-Victorian'.[24] However, some local boatmen, seeing how Grey fought his fish, realised that the existing methods of fighting big game fish were too passive and became converts.

Grey's frank opinions and comments on the trout fishing scene at Tokaanu sparked the fly-fishing versus spoon fishing argument that later resulted in a change to the regulations for Tongariro fishing. He also felt the need for conservation measures, and in 1927, on his second visit, urged that New Zealand anglers form an Izaak Walton League, a nature conservation group which had been formed in the United States whose membership of hunters, anglers and outdoors enthusiasts are active in wildlife conservation and sustainability management.[25]

Zane Grey must have later regretted the controversy he had created with his criticisms of New Zealanders' methods for catching sharks and marlin. In March 1932 he wrote to Trevor Withers, a member of the Auckland Acclimatisation Society, completely retracting his remarks about crude tackle and unsporting techniques, and apologising for them. Withers acknowledged that Grey had taught the New Zealanders much about fighting big game fish, and noted the general acceptance of single hooks in place of treble hooks. Grey's apology, he said, would be appreciated by the swordfishing fraternity, and there would be a warm welcome when Grey next visited New Zealand.[26]

Grey's publicity efforts

The entire tone of the 1926 trip was set by the government's official invitation to sample the fishing and write about it. This meant Grey had to deal with New Zealand tourism officials, publicity men and photographers. He was diligent to the end in his assumed role of distinguished guest, meeting the New Zealand Prime Minister Gordon Coates before sailing home in late April 1926.

He also kept his part of the publicity bargain he had made with the New Zealand government. Upon his return to his home at Avalon on Santa Catalina Island, California, he immediately set to work on a book describing his New Zealand adventures. *Tales of the Angler's El Dorado* was released later in 1926, and, as Grey noted, helped to justify the claims made as to the quality of New Zealand scenery and its fishing. Although the book is devoted mostly to big game fishing in the Bay of Islands, its last chapters describe Grey's and Mitchell's experiences at the Waihora River mouth and then at the Tongariro. The book ends with Grey enthralled by the beauty of the Dreadnought Pool after catching some of its magnificent trout.

Grey's other major publicity effort was a movie film recorded on a later visit. *South Sea Adventures*, released in 1932, contains generous coverage of New Zealand's shark and marlin fishing, but trout fishing on the Tongariro has only a small part in the film.[27]

Grey's second visit — 1927

Grey returned to New Zealand three more times for trout fishing purposes; once in 1927[28] (some 42 days on the Tongariro, after deep sea fishing in the Bay of Islands), again in 1929, and finally in 1932–33.[29] His son Romer Grey and brother RC (Romer Carl) accompanied him on the second visit, with Romer catching a magnificent Tongariro rainbow weighing 15¼lb.[30]

Zane Grey's first visit had opened his eyes to the enormous opportunities that New Zealand had to offer for outdoor recreation. Then, he had spent only a few days days on the Tongariro River, and had not seen it at its best. Once Grey realised that one trip was not going to be enough to sample the many waters available, he made plans to return in 1927 on his own vessel for three or four months of deep sea fishing. The Tongariro rainbows were also on his mind:

> I never had any greater sporting thrill than when I hooked those big rainbows on the six-ounce rod — they would take one hundred yards of line in one run.[31]

Grey's three-masted auxiliary schooner, *The Fisherman*, had been specially fitted out for this second excursion, complete with a glass-bottomed boat and a tank in which to keep fish specimens alive. The vessel berthed at Auckland ahead of Grey and his friends, who arrived in Auckland on 20 January 1927 by steamer, in the company of other American anglers.[32] For this trip, others in Zane Grey's party included his son Romer, Romer's friend John Shields, both aged 17 years, and 'private secretary' Mildred Smith.[33]

This second visit was planned as a much grander affair, with Grey hoping to have two marine scientists join him on a Pacific cruise, while his cameraman Thomas Middleton would film Grey's catches. This time, a month would be allocated to fishing the Tongariro.

The group used their old camp at Kowhai Flat, where the Duke and Duchess of York had stayed some months before. The Americans immediately made their presence felt, but not in a way other anglers welcomed. A profusion of tent ropes and stays spreading close to the river's edge meant outsiders could not progress up- or downstream without passing through the Zane Grey camp.[34]

The 1927 trip was as much a scientific as a sport fishing expedition. Operating from his new schooner, Grey intended to explore not just the fish species which might exist in Pacific waters, but also the marine environment in which they lived.[35]

Perhaps mindful of the single-hook versus treble-hook arguments from the previous visit, Captain Mitchell was careful to offer an olive branch on arrival. All fisherman, he said, were welcome to visit their vessel to view the group's fishing tackle and to learn of any angling methods that would be new to them.

Spoon fishing controversy continues

Grey's 1927 visit coincided with the Easter holidays, when the number of anglers on the river was estimated at 500.[36] The spoon fishing techniques of New Zealand anglers on the Tongariro also made things difficult for fly fishers. Unsophisticated, heavy-tackle enthusiasts hurling ironmongery, complete with treble hooks, across the Tongariro pools (and sometimes across fly fishermen covering the same water) were distinctly unwelcome; Grey thought the Tongariro was too clear a river for such clumsy methods. Spoon fishing disturbed the trout and ruined the chances of fly fishers who looked to fish the same pools.

After completing his second visit to Taupō waters, Zane Grey and his companions confirmed their views that on the Tongariro River spoon fishing spoiled the fishing and should not be allowed. 'It is far from sportsmanlike from the point of view of the American or English angler,' said Captain Laurie Mitchell.[37] While the New Zealand government had advertised the Tongariro River widely overseas,

John Shields, friend of Romer Grey, with a rainbow trout taken from the Tongariro River during Zane Grey's second visit in 1927.

Sir George Grey Special Collections, Auckland Libraries, 1370-U053-12

foreign anglers arriving in New Zealand would be dismayed to find angling conditions vastly different from those to which they were accustomed on their more exclusive home waters.

But under the current regulations,[38] the New Zealand anglers were quite within their rights to fish with spinning tackle, because only the Tongariro River tributaries were reserved for fly fishing. The two methods were ultimately incompatible unless the spoon practitioners were obliging — as only a few were. Further, the regulations did not specify how many anglers could fish in one pool at the same time — a matter, even today, left to angling etiquette — and predictably there were arguments.

Ultimately, Grey became resigned, if not reconciled, to the spoon fishers' presence,[39] conceding that others could love the river, too. However, this meant he was forced to seek out pools where he could fish undisturbed. He and Mitchell found new fishing spots by cutting tracks through the riverside vegetation or using a small boat in the upper reaches to gain access to parts of the riverbank not accessible on foot. Romer Grey, being younger and fitter, used a skiff to get to a pool from which he caught his largest trout of the 1927 expedition — 15¼lb.[40]

By 1927 the extraordinary weight and condition of the Tongariro rainbows were well-known to overseas anglers, no doubt in part courtesy of the New Zealand tourism officials who broadcast the river's fame in the United Kingdom and the United States. As a result, Zane Grey found that he was not the only overseas angler in Tokaanu. Further down river, at Hut Camp, two highly experienced Californian fishermen were making an indelible impression on the locals with their light tackle. J. Cox Webb and the 6 foot 4 inch Frederick K. Burnham caused a sensation when they showed what could be done with their rods of 9½ feet, weighing 6¼oz.

These rods were much smaller than the 14-foot, 20oz monstrosities favoured by New Zealand anglers, but the visitors, in terms of the distance they could cast their flies, were in no way disadvantaged. How the locals must have laboured under the weight of their larger tackle, even when it was used double-handed, as no doubt most of them were! The lesson took some time to sink in; but after the kindly Americans had given some tuition in the use of their lighter implements, there were several converts. Some New Zealand anglers even requested Webb and Burnham to arrange light rods for them.[41]

After almost a month on the Tongariro, with most of their time spent in the Log Pool and Downs Pool, Webb and his friend had scored an impressive tally, averaging eight fish per day at an average weight of 7½lb.[42] They were experienced anglers who quickly adapted to the river's boulder and shingle riverbed, which made it easier to land the large Taupō rainbows within 10 minutes or so. The American rivers, with high banks, often made for a 30-minute battle with a trout. The flies they used were Turkey and Red, Golden Demon, the Matuka and also some salmon flies — Thunder and Lightning and Jock Scott.

> 'The Tongariro River is without a doubt the most beautiful and satisfying river in the world for trout fishermen' said Webb. 'Such quantities of giant rainbows I never hoped to find.'[43]

At the close of Zane Grey's second visit, the gloss was wearing thin. He vowed never to return to the Tongariro because the crowds of spoon fishermen had ruined the river for him. Unlike British salmon rivers, there were no exclusive pools allocated to one rod only. The price extracted from anglers on the Tongariro in return for the system of open water was massive overcrowding at popular holiday times. Surprisingly, in spite of his frustrated protestations, Grey did come back twice, although New Zealand's deep sea fishing was the main lure for him.

Zane Grey's rumoured land purchase

The story persists that Grey was keen to buy some of the riverbank to secure the best water for himself. A number of writers in books and journals have stated that Zane Grey wanted to buy land alongside the Tongariro River pools so he could control the fishing there. Bryn Hammond, writing in *The New Zealand Encyclopaedia of Fly Fishing*, states that Grey had negotiations with the family of Hoka Downs, the Māori guide who showed him where to fish on the upper river.[44] It seems Grey wanted to build a fishing lodge which only he and his friends could use. But because there is currently no direct evidence to substantiate these sorts of assertions, the truth will likely never be known.

Famous people generate enormous publicity, and Zane Grey was certainly one of the most talked-about authors of his day. Such celebrities also attract much speculation among their fans. The land-purchase rumour could be one of those popular tales that start with idle talk, and then, if repeated often enough, slowly become accepted as fact. Something of the kind occurred in 1932 when Gordon Coates, then Minister of Public Works in the coalition government, allegedly told a group of unemployed enduring the Depression to eat grass.[45]

Looking back at those times, we can see that Zane Grey arrived in New Zealand in 1926 with a reputation as one of the wealthiest writers in the world. He had sufficient funds to mount a fishing excursion on the other side of the world lasting several months. His fishing exploits were noted in all the daily newspapers. The New Zealand government, which had invited Grey to tour, also helped to smooth his path by providing free rail travel and allocating staff from the Department of Tourist and Health Resorts to his group. When in Wellington he met with the prime minister, who offered him the use of a government motor car. On returning for his second fishing expedition in 1927, Grey used his massive three-masted schooner, *The Fisherman*, fitted out with all manner of angling equipment.

At this time New Zealand's economic fortunes were declining. Prices for dairy exports to Great Britain, the country's main market, had dropped in line with the downturn taking hold there. Once Grey expressed his great enthusiasm for the Tongariro River and its trout, it was perhaps natural for the seeds of angling xenophobia to take root, with New Zealanders wondering if this wealthy man might take part of their river from them for use as a private fishing preserve. However, in letters to his wife Grey makes no mention of buying land on the Tongariro. And Dolly Grey would need to be told about any such transaction because she paid the bills. Yet a few years later Zane Grey did buy land in Tahiti to use as a base for big game fishing.

One possible source of the New Zealand rumour is the comment that Prime Minister J. G. Coates made in 1926 at the time the government was negotiating with Ngāti Tūwharetoa to secure angler access rights to the Lake Taupō waterside and the margins of rivers. Coates stated in Parliament that the transfer of rights to the Crown was necessary because of the risk that the land adjacent to the Taupō rivers would be leased to foreigners, curtailing the rights of New Zealand anglers.[46] Grey was not mentioned specifically, but his visit was well known to many New Zealanders. Alternatively, speculation may have started based on Grey's obvious wealth from writing.

Grey may have considered the idea of buying a New Zealand block — certainly, at the time of his first visit, funding the purchase would not have caused difficulty. The matter was probably never more than idle fancy. And there are a number of reasons why it would have been impractical anyway. However, by 1927 he had met with opposition from some New Zealand anglers, who regarded him and his party as rich outsiders eager to steal the best fishing for themselves. Possibly that antagonistic climate, coupled with the enormous time and effort required to travel to New Zealand from the United States, weighed heavily against the idea and persuaded him instead to settle on a fishing base in Tahiti.

After making his first trip to the Tongariro, he had bought land at Winkle Bar on Oregon's Rogue River,[47] which gave him fine steelhead fishing remarkably similar to that of the Tongariro. Why travel around the world for months when fishing of comparable quality was available in one's own country?

Grey's third and fourth visits — 1928–29 and 1932–33

Grey's third fishing trip to New Zealand followed the pattern he had established previously: big game fishing in the Bay of Islands for three months (this time at Mercury Island), followed by a journey to Tokaanu for trout fishing on the Tongariro.[48]

By this time Mitchell and Grey had fallen out. In later letters to his wife in 1931, Grey revealed that he was dissatisfied with the way Mitchell had refused to stay within sight when the two were big game fishing at Russell — Mitchell had not followed orders. (Matters were no doubt not helped with Mitchell effectively usurping his friend's role of big game angler by catching a world-record black marlin of 976lb.) Relations between the two men cooled, with Grey making comments to which Mitchell took exception.

The matter flared up once the group left New Zealand for the Pacific Islands. Mitchell consulted a local lawyer in Papeete to sue for outstanding wages. The dispute was discreetly settled by negotiation, a result Grey was very anxious to achieve, because Mitchell was aware of the presence of Grey's mistresses on their fishing trips and could easily have destroyed the writer's public reputation as a happily married man.[49] No publisher would touch an author tainted by such scandal. The man Grey had once described as a loyal fishing comrade, he now called stupid and bull-headed, a liar and a thief. Mitchell remained in Tahiti for a period, and then returned to the United States. It was probably a relief to Grey when Mitchell died in Catalina in August 1931.[50]

As it turned out, all of Grey's plans were undermined by the onset of the economic slump that would become known as the Great Depression, with his own income greatly reduced in 1930. Then, to add to the pressure, the following year he found himself in a dispute with the United States Inland Revenue Service, which claimed he owed $400,000 in unpaid income tax. Coming after the acrimonious split with Captain Mitchell, it made the early 1930s a stressful and unhappy time for Grey, and any visits to New Zealand had to wait.

Grey did return to New Zealand for what would be his fourth and final fishing trip, lasting several months, in early December 1932. While sharks and marlin were the main attraction, he found time for Tongariro fishing in mid-February 1933.[51]

Tahitian appeal

Although thrilled with the angling opportunities New Zealand offered, as mentioned it was Tahiti which ultimately seduced Grey. In 1927 he bought 8 acres of land there and set up a fishing camp. Before their falling out, Mitchell had the job of running the camp when Grey was not in Tahiti. Already aware that the wild America he loved was changing, Grey was enchanted by the French Polynesian region's tropical beauty and its wild remoteness, although the hot and oppressive climate did not suit him and made the task of writing difficult.

In later years Grey ordered a 46-foot fishing launch, the *Frangipani*, to be built by the Auckland firm Collings and Bell. His previous custom-built boat, *Fisherman II*, had been built in the United States. It proved a total disaster; hugely expensive and with the capacity to roll in even the smallest waves. But the *Frangipani* passed the test. Grey collected it in December 1932 on his fourth and final visit to New Zealand, took it to the Bay of Islands, and then had his men sail it to Tahiti for more fishing adventures.

Later adventures

In the mid-1930s, Grey became interested in the steelhead fisheries of Oregon. He fished often for steelhead rainbow trout in the Rogue River. Later, he moved on and fished the North Umpqua, a little to the north of the Rogue. The Umpqua yielded rainbows every bit as large as the Tongariro. Always seeking new fishing grounds, Grey discovered Australia's Great Barrier Reef in the mid-1930s. Once he experienced the long coastline, populated by giant white pointer sharks and other exotic species, he switched his angling allegiance. In August 1936 he was describing Australia as the up-and-coming big game country for anglers, and one which would prove to have the finest deep sea fishing in the world.[52] New Zealand thereafter saw him only as a passenger in transit.

In reflecting on his restlessness, Grey once said that 'in the fever and turmoil of civilised pursuits, I was haunted by wild places. It was simply the longing to go back to some former stage of life's development.'[53]

Grey's writing

Although largely allergic to reviewing and rewriting his work, Grey was adept at turning out a sound first draft of a story. He took a highly disciplined approach to writing, rising at an early hour on fishing trips to write an account of the previous day's events. He was a highly observant man, keenly aware of his reactions to his environment, and capable of expressing his emotions with an eloquent simplicity. And as noted earlier, Grey channelled his manuscripts through his wife and editor Dolly, who ensured the quality of his published work.

His writing reflected his life as an adventurer and explorer. The Tongariro fitted well with his idea of what a trout river should be: wild, brawling water, more than a little dangerous, with large trout to test the skill of the most experienced angler. Grey's writing, therefore, is a report on experience, but it goes beyond mere description to become a rich angling mosaic. It has a freshness which may have emanated from his habit of composing notes at the waterside for later completion. If so, that habit allowed the spirit of the Tongariro to become a part of his work.

Notes made during his 1929 expedition to the Tongariro show how Grey draws the reader into the scene. He describes wading into the river, hearing the sound of rapids, the shock of ice-cold water, the scent of campfire smoke, the sunset and the moon in the sky.[54] That said, Grey was also self-centred in his writing. His work lacks the humility of Arthur Ransome or the self-effacing humour of Negley Farson. There is some similarity to Ernest Hemingway's *Green Hills of Africa*, with its focus on the hunter's need to pursue and overcome his quarry.[55]

Zane Grey continued to praise the quality of New Zealand fishing in overseas journals whenever possible. An article published in March 1932, 'Most Famous and Beautiful Trout Stream in the World', tells of the fine rainbow trout available, and how Fred Burnham used superb casting to take a limit from the Log Pool. Local fishermen, awed by the spectacle, retired to the riverbank to watch.[56]

Grey's description of the Tongariro evokes impressions of a cathedral:

> The beauty of the Tongariro is on a par with its fishing possibilities. The pumice stone cliffs resembled painted walls and their reflection in the magnifying crystal waters resembled painted windows. And the foliage along this river is of so rich a dark, deep, shiny green, the ti forests so fine of texture and so colourfully white and pink with tiny blossoms, as

> lovely as the chamice of Catalina, and the dense copses of giant ferns, lacy as pine-needles on the bough, all seemed to belong protectingly to the Tongariro alone, to hide its treasures of emerald pools and aquamarine rapids, to preserve its beauty and its rainbows from the greed of sportsmen.
>
> Through this stretch of tangled forests it winds along in an utter solitude, broken only by the melody of the bellbird and the tui, and the mellow roar of rapids. And at last it slides out into the open, to linger in foam-flecked pools and spread wide over bars and split around green islands and run deep and black under sheer gray walls, at last to weary of restraint and rush madly down, a turbulent green-white spirit-forsaken river.[57]

He closes by urging all fly fishers to get to the Tongariro and experience its charms.

This complex and contradictory man died in California on 23 October 1939. While he garnered both fame and fortune, his eternal quest for ever greater challenges in the outdoors extracted a heavy price from his wife and family.

CHAPTER 8:

Royal visitors

The Duke and Duchess of York will arrive at the camp prepared for them at Kowhai Flat, on the banks of the Tongariro River, this evening, and will spend all day tomorrow and Wednesday trout fishing.

— *THE NEW ZEALAND HERALD*, 27 FEBRUARY 1927

THE WORLDWIDE FAME OF THE TONGARIRO'S RAINBOWS received a huge boost in the late 1920s when the Duke and Duchess of York toured New Zealand. Their itinerary was a busy one, with many visits to small towns and personal contact with the public, but fishing was to play a significant part. The tour started with big game fishing in the Bay of Islands before the couple made their way south to Rotorua, centre of the thermal district. From there, it was on to Tokaanu for a two-day rest period with the bonus of trout fishing.

All along the road on Lake Taupō's eastern shore local residents turned out for a sight of the royal party. Some decorated the roadside with ferns, but at the Waitahanui river bridge one group decided to show what the fishing prospects were. As the cars of the Duke's party — 25 in all — crossed the bridge, 20 anglers formed lines on each side and held their fly rods to form an archway of honour. Trout recently caught lay at their feet as proof of the fishery's bounty.

A riverside campsite

Camping outdoors was literally the order of the day it seemed, as the Duchess of York had rejected a suggestion of 'more regal' accommodation. In reality, the only accommodation in the area was the

Future king at play: the Duke of York fishes the Tongariro River in 1927 under the watchful eye of his 'gillie', Conservator of Fish and Game Frederick, Moorhouse. Gaffs were legal in those days.
Auckland Libraries Heritage Collections 1370-351-1

Tokaanu Hotel, which was comfortable enough for most tourists, but had not been built with visiting aristocrats in mind. If the Duke and Duchess (and staff) had lodged there, they would have had to commandeer the entire premises. Government officials probably decided to make a virtue out of necessity and set up a riverside camp, thereby saving travelling time and allowing privacy. And indeed the campsite was a worthy forerunner to our much-beloved modern-day 'glamping' site. For, even though the Duke and Duchess might be sleeping under canvas, the tents had wooden floors and all conveniences were supplied, from an electricity generator and arc lights to catering by a government chef. As this part of the tour was intended to give the visitors a break from the formality of the tour, only their immediate staff members (maids included, of course) accompanied them.

The Duke and Duchess arrived late in the evening of 28 February 1927, weary from days of official duties. They settled into camp, relaxed by the log fire, and went to sleep lulled by the mighty roar of the rushing Tongariro waters a short distance away. They rose early the next morning for fishing. To ensure success, the government had engaged local experts to put their guests on to the fish. No eminence less than the Conservator of Fish and Game, Frederick Moorhouse, guided the Duke of York, while the Duchess had the counsel of one Captain Knowles from Rotorua.

Low water conditions at the campsite for the royal couple in 1927. Kowhai Flat appears at upper left.

Archives New Zealand, Archway Item ID:R21010253, Archway Series Number:6539

Royal catches

Unfortunately, February on the Tongariro was not an ideal time of year for trout fishing. The runs from the lake had not begun, and few fish were being taken from the upper pools. Despite the challenging conditions, the visitors got into waders and caught fish. The Duke opened his account with five trout, one weighing 8½lb,[1] but fishing being what it is his staff had better luck. Naval equerry Lieutenant-Colonel Buist caught seven fish, while the Earl of Cavan played one fish thought to be 14lb for almost an hour, only to lose it when it was about to be gaffed. He was left with the consolation of a good tale of the one that got away. The Duchess of York joined the Duke and fished for most of the day in sunny conditions, but had no takes.

On day two, the Duke was out early, eager to add to his total, but met with no success. He journeyed down from his camp to the Log Pool, only to find that other anglers, up even earlier, had fished his chosen pool already. When he arrived, they insisted on leaving the water to him and did so, despite his generous invitation for them to remain.[2] His own ensuing disappointment was cancelled out by the success of his wife, who caught a fine 7lb rainbow, to the general delight of the party.

On this second day the Duke had risen early to make the most of the short time available. However, he still had time to good-naturedly

The Duchess of York casting in the Tongariro River. Captain Knowles of Rotorua is at her side, while a press photographer lurks in the background.
Auckland Libraries Heritage Collections 1370-351-4

accept the attentions of the press photographers, who had been allowed on the riverbank to interview him. Arriving at 7am, the journalists found the Duke already fishing the Log Pool from the other side of the river, and he crossed over to talk to them. Dressed in waders, felt hat and jersey, he was indistinguishable from any other angler, and, after showing his casting skills for a time, he asked if they would like to meet the Duchess, who was fishing nearby. Naturally they did, and when HRH appeared she, too, had time for an informal chat about the fishing.

It was on the Log Pool that the Duke suffered a mishap. After hooking a nice rainbow of about 7lb, he drew it into shallow water. The gillie (presumably Frederick Moorhouse) waded into the river and made a mighty swipe with his gaff, but missed the fish and broke the leader. The trout swam off. As the Duchess of York was not present, the Duke could tell Moorhouse in sailor's language exactly what he thought of that little effort.[3]

Although it was brief, the visit by the future king and queen received good coverage in the British press and gave an added boost to the Tongariro's reputation as a world-class trout fishery.

CHAPTER 9:

A public fishery emerges

The Prime Minister, Rt Hon. J. G. Coates, in his capacity as Native Minister, will visit Waihi, Lake Taupo, on April 21 to negotiate with the natives regarding fishing rights in Taupo waters.

— ***THE MANAWATU STANDARD*, 31 MARCH 1926**

EUROPEANS SETTLING IN NEW ZEALAND in the nineteenth century had an eye on sport fish acclimatisation from the get-go. In 1867 the government enacted the Protection of Animals Act to provide the necessary legal framework, and to legitimise the activities of acclimatisation societies in liberating game, such as pheasants, hares, quail and duck. The Salmon and Trout Act 1867 was passed in the same year, even though neither salmon nor trout had yet been successfully established for sport fishing purposes. The Act allowed the government to issue regulations dealing with trout and salmon propagation, their habitat, and restrictions on fishing methods. It was also in 1867 that the Canterbury Acclimatisation Society received brown trout ova from its Tasmanian counterpart. The fingerlings reared from this shipment were liberated locally and their progeny used to spread trout to other parts of the country.

Nearly 20 years later, in 1886, the government issued regulations protecting introduced fish species throughout New Zealand.[1] The Taupō fishery, including the Tongariro River, formed part of the Rotorua acclimatisation district, an area administered by the Auckland Acclimatisation Society.

The fishing season ran from 1 November to 30 April (spring, summer and autumn). As a general rule, the permitted lures or baits were restricted to natural or artificial flies, insects or fish used on rod and running line. The minimum size for trout was 10 inches in length, although the daily catch limit was measured in the weight of fish (30lb), not the number of fish.

The government takeover

Fish and game laws applied on a national level and were overseen by acclimatisation societies in different parts of the country. At the beginning of the twentieth century the Auckland Acclimatisation Society issued licences for the Rotorua and Taupō regions, but, as we have seen, this was not to last. The outstanding quality of the fishing in the thermal regions proved too attractive for the government to resist. In 1907 Parliament passed laws to take the fisheries — and with them the income from fishing licence sales — away from the Auckland society's control.

This autocratic and arbitrary measure was hotly contested at the time, but the government could see the enormous possibilities for Rotorua as a tourist and health centre and would not be swayed, even though a public petition against the measure gained the support of a parliamentary committee which heard submissions on the matter.[2] The result was a new acclimatisation district which included Lake Rotorua, Lake Taupō and the rivers feeding them.[3]

As we learnt earlier, at first the Department of Tourist and Health Resorts administered the Rotorua acclimatisation district, but with its staff not being fisheries managers, the quality of the fish declined.[4] When this problem emerged around 1912 a restructuring followed; the Department of Tourist and Health Resorts issued the licences to take fish and game at Rotorua and Taupō, while the Department of Internal Affairs, newly appointed in 1913, looked after the science of fisheries management.[5]

The new fishing season at Taupō lasted from 1 November to the following 31 May; an extension of one month. These dates were retained in the regulations made in 1926 for the Taupō fishery, and continued to apply until the closing date was brought forward to 14 May in 1939.[6]

A whole-season licence cost £1, with reduced fees of 5 shillings for women and for boys under 16 years. From 1 February each year, a reduced fee of 12 shillings 6 pence was charged for licences that would expire on the following 31 May.[7] Licences could also be purchased for periods of a month, a week or one day.

When compared with future regulations, the 1914 rules were rudimentary. In terms of lures, anglers were restricted to:

> . . . the natural or artificial fly, natural or artificial minnow, any small indigenous fish, insect, grasshopper, beetle or spider, and any form of spoon bait . . .[8]

provided it was attached to a rod and reel. Kōura (freshwater crayfish), worms, creepers and native grubs (huhu or matai) were specifically excluded.

Spinning and trolling from a boat were also unlawful in any of the rivers and streams flowing into a number of specified lakes in the Rotorua acclimatisation district, 'all of such being reserved for fly fishing only'. At first Lake Taupō and its rivers were part of this exclusion, but after several months the Taupō region was removed from the protected areas.[9] This meant that any lawful angling method could be used by licence holders in the Tongariro River and its tributary streams. However, the regulations went further and made specific provision for artificial fly to be the only method allowed in certain waters, including the Waitahanui River on Lake Taupō's north-eastern shores.[10]

There were no restrictions on the number of trout an angler could kill in one day, nor on the hours of fishing. That meant anglers with sufficient stamina and persistence could fish around the clock.[11]

Taupō waters agreement initiated

The regulations made in 1914 continued largely unchanged until the mid-1920s, when it became apparent to the government that some Māori landowners in the thermal regions were charging anglers for the right to cross their land. One enterprising group of Māori in the Urewera district had even formed their own trout fishing club, and required all outsiders to pay a 5-shilling 'membership fee' before being allowed to fish in local rivers.[12] Access charges also occurred on the Waitahanui, a popular trout river to the south of Taupō town, and at Tokaanu. But tourists who had bought £1 licences giving them the right to fish throughout New Zealand objected to paying another fee to cross Māori land alongside Taupō rivers.

These complaints made the government feel obliged to act. In 1924 it passed legislation authorising the Native Minister to negotiate with those people claiming to be owners of land bordering 'Taupo waters' — that is, Lake Taupō, all rivers and streams flowing into it, and the Waikato River down to the Huka Falls. Negotiations would cover fishing rights in Taupō waters and the riverbeds and their margins. The Crown was also to have the right to charge for fishing licences, with a portion of the fees '[to] be appropriated to the Natives interested and . . . distributed among them or applied for their benefit . . .'[13] Any agreement reached could be put into effect provided the Native Minister was satisfied a substantial majority of Māori persons present approved.

Tourism officials were also keen to see Taupō get separate treatment. The law allowed a district acclimatisation society to issue a fishing licence authorising the holder to fish anywhere in New Zealand. Fishing was not restricted to the region controlled by the society that issued the licence. This loophole meant an angler who bought a fishing licence from the Auckland Acclimatisation Society could fish at Rotorua or Taupō without paying a further fee to the Department of Tourist and Health Resorts, which incurred the cost of stocking and maintaining the lakes and rivers in those areas.[14]

Negotiations continue at Waihi meeting

Talks went on for more than two years between the Crown and Māori before the parties settled terms. On 21 April 1926, Prime Minister Joseph Gordon Coates travelled to Waihi village at the southern end of Lake Taupō hoping to conclude matters. Coates was a Northland farmer who had entered Parliament in 1911, cannily choosing to run in the Northland electorate of Kaipara as a Liberal Independent rather than immediately throw his weight behind the Liberal government. By 1914 he had followed the example of other independents and joined the Reform government of William Massey. Following army service in France in World War I, he was given a ministerial post in 1919 at age 41 years, making him the youngest minister in the administration.

Coates had a relaxed, informal manner which made it easy for others to feel comfortable in his presence. Having grown up around Māori, he had got to know them, learnt some te reo Māori and, most importantly, come to like them. As one historian notes, Coates felt completely at

Rt Hon. Joseph Gordon Coates, Minister of Native Affairs, who negotiated the agreement over fishing rights in Lake Taupō.
Alexander Turnbull Library, Wellington, New Zealand. /records/23085196

home in their company.[15] He was the natural choice as Native Affairs Minister when he was appointed to that office in 1921. He later became prime minister, in 1925, whereupon he was dubbed 'the jazz Premier'.

The Waihi fishery congress was a big event. Many people attended the meeting, if the catering arrangements laid on by the government are any indication: 900 loaves of bread, 2,000lb of beef and 1 ton of potatoes provided by the local prison.

Coates handled the negotiations skilfully. Ngāti Tūwharetoa wanted £15,000 paid to them annually in return for all their rights over the lake waters. This offer was broadly similar to the agreement the government had reached with Māori over the thermal lakes in the Rotorua district. Coates dismissed this sum as extravagant.[16] The government, he said, was interested not in ownership of the lake but in doing a deal on fishing rights in Lake Taupō. In exchange for those rights he offered one-half of the government income to come from increased fishing licence fees in the Taupō fishery.

After a well-timed break for lunch, the Māori contingent conferred to consider their position. Paramount chief Hoani Te Heu Heu broke the deadlock when he proposed a resolution that the tribe should cede to the Crown its fishing rights in and over Lake Taupō in return for an annual payment of £3,000 and a further top-up payment of 50 per cent of all licence fee payments exceeding that sum. The resolution was carried without objection and conveyed to Coates, who accepted the offer in principle, with the details to be settled later on. The question of fishing rights in Lake Taupō's rivers was also marked for further discussion.[17] However, the government's failure to clarify which rivers and streams it was concerned with was to cause problems in the future.

Not all the Māori present were happy with the proposals. Those owning land alongside the Waitahanui River did not wish to see any change. They were content to carry on farming the fishery, as one frustrated government official commented,[18] and obtaining regular income (said to be 2 shillings and 6 pence per day) from anglers.[19] But Te Heu Heu used his undoubted rank to speak for them all.

Having made significant progress on the fishing rights issue, Coates returned to Wellington to deal with more mundane political matters. Back at Waihi village, one matter came up late in the day. This was the effect of the proposed agreement on Europeans who had private rights over Māori land bordering Lake Taupō rivers. The only applicant was the estate of Robert Jones, the lessee of land beside the Tongariro River on which he had set up the three fishing camps, discussed in more detail below. As we have seen, when Jones died in 1924, the estate was represented by his wife's brother, John Asher. The day after Coates returned to Wellington, there was an inquiry into this question before Judge Browne. Asher was reluctant to disclose financial details of Jones's Tongariro fishing rights, although it seems that leases had been granted for 21 years with the money paid in advance.

Taupō waters agreement concluded

In late July 1926, representatives of Ngāti Tūwharetoa travelled to Wellington to discuss the outstanding issues. Coates met them again, and this time enough of the main points were settled to allow an agreement to be reached. Chief Hoani Te Heu Heu signed on behalf of the iwi and the prime minister's private secretary signed for the Crown.

The agreement gave title to the beds of all Taupō waters to the Crown as a public reserve, with public access reserved around the lake bed for 1 chain (22 yards/20metres). 'Taupō waters' were defined as the lake and all rivers and streams entering it, together with the first few miles of the Waikato River, of which Lake Taupō is the source, ending at the Huka Falls a short distance to the north.

The holders of 'special licences' (fishing licences) were allowed foot access over strips of land 1 chain wide on each of the Taupō rivers and streams. The rivers to which this right would apply were to be specified by government order at a later date. This selective approach implied that the government did not intend every river to be covered. Its main aim was to ensure unfettered fishing access to the more popular rivers, such as the Tongariro and Waitahanui.

Islands situated in the lake were reserved to Māori — a status that remains to this day. Motutaiko, near Lake Taupō's south-eastern shore, is the chief example. The island has great significance for Ngāti Tūwharetoa because of its use as a burial ground for ancestors.

By this time, word had got out about the compensation sought for Robert Jones's fishing camps. Other riparian owners said they, too, had been earning income from fishing and camping fees, so a clause in the agreement provided for a special tribunal to assess these claims.

The next step was to give the agreement the force of law by Act of Parliament. In debating the legislation in the House, the prime minister noted that the question of the title to the bed of Lake Taupō had been an unresolved issue for more than 50 years. The local Tūwharetoa iwi had consistently based its claim to the lake on the Treaty of Waitangi, while the Crown had asserted it was the lawful owner.

The government's goal was to remove the problem posed by Taupō landowners charging for fishing access. Interestingly, when debating in Parliament Coates did not refer to this directly, but chose instead to raise the threat of a takeover by foreigners with a consequent loss of fishing access:

> . . . there has been a danger of the banks of the rivers being handed over on lease to men outside the Dominion. Quite a number of negotiations had actually taken place, and it is not a stretch of imagination to say that in the course of a few years the banks of these rivers would have been owned by foreigners, people who were not New-Zealanders at all.[20]

In negotiating the agreement made with Ngāti Tūwharetoa earlier in 1926, Coates had taken an entirely different tack, telling the assembled iwi members at Waihi village that the Crown was not concerned with ownership of Lake Taupō but with ensuring that Māori received some financial benefit from the burgeoning trout fishery.[21]

Access to the Taupō rivers for all anglers was a major objective for the government. With the lower Tongariro riverbank under the control of private fishing camps like those run by Robert Jones, and few pools available in the upper river, much water was out of bounds to the ordinary angler who was not a tourist. The Tongariro River was a de facto private fishery.

Jones had paid local Māori landowners for the exclusive right to set up anglers camps on the lower Tongariro River (£90 per year, according to some reports), and these camps were thought to bring him £340 per week when fully occupied during the fishing season.[22] And because accommodation was limited, as we have seen there was no guarantee that a place would be available; one South Island visitor to Tokaanu found that the camps were fully booked from the beginning of February 1927 to the end of the following May.[23]

Campsites on the eastern shore of Lake Taupō were also controlled by local identities like Captain Thomas Ryan, who paid £80 to the local Māori landowners for these rights. The statutory right of riverbank access was designed to overcome this barrier.[24]

These longstanding agreements were threatened by the government's intervention in the form of the Taupō waters agreement, which was given legal effect by an Act of Parliament that came into force on 11 September 1926.[25] Unfortunately, as discussed below, the law as enacted failed to reflect the terms of the agreement reached with the iwi, let alone these other informal European stakeholders.

Government proclamation

A second legislative stage started in October 1926 with a government proclamation under the new Act.[26] An access strip 20 metres wide was given the status of Crown land on each side of the Tongariro River from the delta to a point some 14 kilometres upstream at the junction with the Whitikau Stream (a point very close to today's upper fishing limit). Holders of fishing licences were granted right of way along the access strip but only on foot — it was not a public road.

An exception was for Jones Island, where Delta Camp was located. This island did not form part of the riverbed vested in the Crown, and was largely excluded from the right of way reserved under Part II of the proclamation. Part III, however, gave anglers right of way over a defined portion of the south-western corner of the island and an access strip around the perimeter of the remaining land, which meant there was foot access to the Tongariro River mouth.

Camping was allowed along the banks of the Tongariro, but only on the western (true left) bank of the river.[27] No camping was allowed on the right bank, access to which was restricted to persons on foot and for purely angling purposes.

All of these access and camping rights were subject to the overriding requirement that they could be enjoyed only by persons holding a fishing licence (or camping permit), although guests of licensed campers could enter onto camp sites as permitted by the regulations. Access to the marginal strip of land was for angling purposes only.[28]

Fishing regulations

Under the 1926 regulations for the newly created Taupō district, fly fishing was the only method allowed in the Tongariro River tributaries, except for the Poutu River. That meant all legal angling techniques, including spoon fishing, could be used on the Tongariro's main stem from the upper river down to the delta.

Under these new rules, river access was no longer tied to the fishing lodges and camps. The first Easter holiday under the new system saw an estimated 200–300 anglers trying their luck. In the scramble for water, as one newspaper put it, fishing etiquette went by the board.[29]

The new rules for Taupō anglers were fine-tuned the following year. Under the 1927 regulations,[30] artificial fly (exclusive of a spinning fly) was the only permitted lure in:

- all tributaries, other than the Poutu River;
- the main stream of the Tongariro from the point where it was joined by the Hatchery Creek down to the mouth; and
- that part of Lake Taupō within 200 yards of the Tongariro River mouth.

The 1927 regulations also contained explicit directions as to how campsites were to be used. It was at this point that permanent campsites were introduced.[31]

Fishing licences and fees for Taupō

Once Taupō became a separate fishery administered by the government, people fishing there had to buy a special fishing licence. Under the 1926 regulations, licences could be bought for a full season, a week or a day. The cost of a licence depended on where the angler resided, with the locals paying the smallest fees.

For example, fees for a full season licence were:

Taupō residents £1 10/-
Other New Zealand residents £3
Overseas residents £6

Women married to Taupō residents and young people paid smaller fees, but for overseas anglers an odd restriction applied at first: they could buy only one daily or weekly licence in a season.[32] This rule created enormous inconvenience for tourist anglers who visited the Taupō district several times in the course of their holidays and were effectively forced to buy a season licence for £6 if they wished to continue fishing for more than eight days. Moreover, the visitors were misled by extensive New Zealand government publicity that a fishing licence covering all waters throughout New Zealand cost £1.

It was not an easy matter for an overseas angler to buy a new licence, because at first there were not many places at Taupō authorised to sell them. The English angler Edward W. Dean summed up the tourist's dilemma:

> . . . how about the overseas man who gets to Taupo, or within a few miles of it, and finds the only place to obtain a licence is at the Taupo Post Office before 4pm? He finds the charge is 15s, or 7½ times as much as for a resident. The next day he goes for another day's licence, and has to pay a further £2, which will entitle him to fish for a week. After he has fished a week he has to pay a further sum of £6 for any further period.[33]

This rule was no doubt intended to maximise licence income for the Crown, but led to considerable ill-will. The government therefore removed it when it adjusted the scale of licence fees the following season, enabling an angler to buy as many day or week licences as he or she wanted.[34]

Fly fishing only rule

The imposition of artificial fly as the sole angling method had major implications for spin fishermen on the lower sections of the Tongariro River. By early 1926 Taupō fly anglers were complaining about the effects that trolling and spin fishing were having.[35] In 1927 the Hawke's Bay Acclimatisation Society lobbied the government for further restrictions on trolling and spin fishing near the mouths of rivers entering Lake Taupō: in effect, for the creation of fly fishing only areas within the Taupō fishery.[36]

Fly fishers considered themselves entitled to some elbow room for their form of sport. One angler supporting the submission, J. H. Edmundson, considered it a sacrilege that spoon fishing was allowed in the mighty Log Pool. He argued that water should be reserved exclusively for fly fishing from the Taupō–Tokaanu road bridge down to the delta mouth.

At that time, a fly angler had to share the river with others who could cast weighted spoons and other lures to every corner of a pool. There were no restrictions on the number of anglers per pool, for the government (perhaps wisely) left that question to be sorted out by the anglers themselves under the accepted rules of fishing etiquette.

Many foreign anglers resented the increased popularity of the fishery and the continued intrusions by spoon fishers who crowded the pools. Some, like Sir Douglas Hall, DSO, Coldstream Guards, accepted the situation and did well with the rod (Sir Douglas banked a 13lb rainbow).[37] Others declared that the Tongariro was finished and vowed never to return,[38] like the Australian who had come to New Zealand for nine years. Angling will always have its share of Jeremiahs.

Some boat fishing was permitted in rivers entering Lake Taupō. The 1927 regulations prohibited fishing with a spoon from a boat, which meant that fly fishing from an anchored boat, or trailing a fly from a moving one (known as harling) remained legal methods. These rules applied in the 1928–29 season, although the fly fishing lobby had one small victory. For the month of May (when the Tongariro runs were usually at their peak) it was unlawful to use an artificial minnow or any form of spoon bait in any river entering Lake Taupō.[39]

The upper sections of the Tongariro River were still permitted to the spin angler, and it remained an effective method of taking trout. At the start of the 1931 season, good bags were reported by anglers fishing with both fly and spoon from the stretch on the upper river between the Dreadnought Pool and the hatchery.[40]

The 1929 regulations also specified that no trolling or harling with a fly was permitted in those parts of the Tongariro River (and other rivers entering Lake Taupō) reserved for fly fishing. Boats continued to travel up and down the lower river as far as Grace's Reach, to the consternation of some anglers who felt the intrusion disturbed the trout.[41]

It is clear from the regulations issued from 1926 to 1929, and later, that there was no confusion in the minds of fishery administrators as to the proper name of the river entering Lake Taupō at Tokaanu. The regulations refer to the Tongariro and Waikato rivers in a manner which indicates they regarded them as separate waterways.

Māori compensation claims

The Act of Parliament[42] implementing the Taupō waters agreement was poorly drafted and failed to reflect the negotiated terms. Prime Minister Coates had never intended that individual Māori landowners could claim compensation for the loss of the beds of the rivers or the imposition of angler access rights on riverside land.[43] Only the more heavily fished rivers — the Tongariro and Waitahanui — were meant to be the subject of compensation, but these were not specifically named. Many Māori spotted the scope for additional compensation and filed claims amounting to more than £60,000.[44]

In 1928 Ngāti Tūwharetoa chief Hoani Te Heu Heu issued legal proceedings in reliance on the compensation rights inadvertently passed into law. Te He Heu argued that he and other Māori owning land bordering Lake Taupō's tributary rivers should be compensated because they had lost ownership of Taupō riverbeds and their rights had been injuriously affected by the anglers' access rights imposed on riparian land.

In summing up the effect of the legislation the court stated:

> . . . the intention expressed in the agreement has not been given effect to. The [section] purports to give to the natives the right to apply for further compensation for matters which they have already agreed to accept the compensation provided for in the agreement; while in the very matter in which it was agreed they should be entitled to claim compensation, no right of compensation has been given.[45]

Faced with an apparently clear compensation right, the Solicitor-General, Arthur Fair, simply told the court that the law had failed to correctly implement the 1926 agreement. Parliament had never intended to provide the claimants (all of whom would share in the annual Crown payment and fishing licence revenue) with double compensation by also granting them sums based on lost fishing rents.

During the hearing, the judges raised section 89 of the Fisheries Act 1908, which made it unlawful for anyone to sell or lease fishing rights. Should compensation be granted for the loss of money earned in an unlawful way? Arthur Fair was true to his name and said the Crown would not seek to rely on such a technical point. Clause 13 of the agreement was intended to give compensation for the funds the Māori landowners had actually been receiving, whatever the true legal position. Prime Minister Coates, when Fair told him of this some days later, approved of it as being in the spirit of the agreement.[46]

In almost all litigation, a court will give a decision in favour of one side or the other, provided it has jurisdiction to do so and the judgment handed down is not otherwise pointless. In Te Heu Heu's case the Full Court had to assume that Parliament meant the words of the legislation to have the meaning that they conveyed; compensation was to be paid. But the law was defective because it was inconsistent with the Taupō waters agreement. In the end the court decided not to rule on the meaning of the law 'to obviate the possibility of injustice to either party by a forced interpretation'. It was, said the judges, a matter for Parliament to remedy.

Judge Ostler's predicament

One of the five judges who heard the claim was very familiar with the Taupō region. Judge Henry Ostler was a keen outdoorsman who had hunted lion and elephant in what was then Tanganyika, where he had a share in a 10,000-acre (4,047-hectare) farm. He had also bought two blocks of land beside the Tongariro River, just upstream from the main road bridge.

In late July 1926, two years before the court hearing, he wrote to the prime minister disclosing his ownership of the Tongariro River property. He urged the government '. . . to prevent any private individual getting a monopoly of the Tongariro fishing and making

The hunting and fishing judge, Henry Hubert Ostler. Should he have recused himself?
Alexander Turnbull Library, Wellington, New Zealand. /records/22613991

a huge profit out of it, by charging prices which can only be paid by rich tourists. This is contrary to the spirit of the country . . .'[47]

No one seems to have thought Ostler's land holdings at Tūrangi disqualified him from adjudicating on the Lake Taupō compensation claim. Prime Minister Coates seems to have ignored the matter. In mid-December 1926 a newspaper report casually stated that Ostler would be fishing with his brother-in-law on the Tongariro River, where the judge had a cottage.[48] One week later the Supreme Court issued its unanimous decision dismissing Te Heu Heu's compensation claim. Ostler wrote a joint judgment for three of the five judges.

It is very likely that nowadays a judge in his position would be required to stand aside from the case. The test is one of apparent bias: would a fair-minded observer who knows the full facts think the judge might not be impartial? However, even had Ostler recused himself it seems the legal issue was insoluble and the result would have been the same.

Final claims settlement

Negotiations between the Crown and the claimants dragged on for years, interrupted by New Zealand's involvement in World War II. Puataata Alfred Grace, who had organised the Tongariro claimants and acted as spokesperson, offered to reduce the amounts sought by one-third in order to reach a settlement, but without success. In the end the Taupo Land Claims Commission fixed the compensation on 13 December 1948 at £45,600. The government approved this sum (which included an interest component) on 20 April 1949.[49] The payment covered only those people owning Taupō riverside land on the date of the original government proclamation in 1926. In making the award the Commission noted that the right of the Māori landowners to reserve to themselves the right of access to fishing waters had 'very considerable value'.[50]

CHAPTER 10:

The name of the river

Duplication in the naming of the most famous trout stream in New Zealand — if not the in world — has been causing a certain measure of confusion.

— *THE EVENING POST*, 28 JULY 1928

TO ANGLERS OF TODAY, the name of the Tongariro River appears both self-evident and entirely natural, given that Tongariro mountain lies to the south and may be presumed to be the place at which the source of the river rises. According to Māori legend, the river is named after Mt Tongariro. 'Tongariro' means 'seized by the cold south wind' and reflects the story of Ngātoroirangi, an early Māori explorer who, being unused to snow, had to call on fire to survive an ascent of the mountain.[1]

It is therefore surprising to find that a long-running dispute over the correct name of the river should have occupied so much time and effort of so many people in the mid-twentieth century. As events showed, anyone could marshal an argument one way or the other, but no one could provide a conclusive answer. Official correspondence reveals that the controversy had more twists and turns than the river itself.

Confusion begins

Government files record the controversy beginning in November 1926, innocently enough, when someone pointed out to the Department of Internal Affairs the possibility for confusion among anglers. The fishing regulations referred to the Tongariro River, while others at times used the name Waikato.

In classic bureaucratic fashion the issue was diverted to the Department of Lands and Survey, which referred it to the Honorary Geographic Board of New Zealand. C. A. (Cecil) Whitney, keen Tongariro angler and future stalwart of the river's angling club, became aware of the issue in mid-1927. He felt that the correct name was the Waikato, because a stream of that name was the ultimate source of the river, which, joined by other streams, found its way into Lake Taupō after passing between the Ruapehu and Kaimanawa ranges. The Tongariro was a mere tributary. As an alternative, he advocated the name 'Ruapehu'.

In late 1927 or early 1928 the Automobile Association (Auckland) erected notices on the main highway bridge spanning the river at Tūrangi, referring to the 'Waikato River'. They were torn down within weeks, probably by local residents opposed to the change. One detail points to the heat which the naming issue was generating — the signs did not simply disappear, but were, according to one report, smashed to pieces. Notably, the Automobile Association signs erected in similar fashion at bridges spanning other eastern Lake Taupō tributaries were not interfered with.

G. F. (Frank) Yerex, Conservator of Fish and Game, was aware of the indignation of anglers and Tūrangi residents when the notices were first erected, noting dryly: 'This may account for their removal.' Yerex based his view on the use of the term Tongariro by local Māori and the fame the river had acquired with overseas anglers.[2]

A little later, an official from the Department of Internal Affairs attended a meeting of the Honorary Geographic Board of New Zealand[3] held on 9 March 1928, and pressed for the use of the name Tongariro, justifying that view because the river was known throughout the fisheries world by that name.

By now, thanks to the eminent visitors Zane Grey and the Duke and Duchess of York, the river known as the Tongariro was well on the way to achieving an international reputation for the quality of its trout fishing. Later that year, the Geographic Board recommended that on maps and other documents the river should be called the Upper Waikato, with 'Tongariro' in parentheses.[4] The Minister of Lands had appointed the board in 1924 to investigate and counsel the government on geographical matters, but it could not itself make any formal decision that would be legally binding.

In March 1929 Whitney made an unsuccessful request to the

Minister of Internal Affairs, the Hon. P. A. de la Perelle, for ratification of the name Waikato over the entire river. He also suggested an alternative name, Awapapa, which he said meant 'the father of the waters'. The minister was unmoved by this approach. Tongariro was the name by which the river was known to all anglers and to others, which made a name change impracticable. A further submission two years later urging the department to adopt the decision of the Honorary Geographic Board of New Zealand and use both Waikato and Tongariro also made no impression. It is clear that by this stage Internal Affairs was well aware of the tourism value of the river called Tongariro, and was not about to make any change that might damage the reputation already created.

Historians argue

In early November 1931 newspaper correspondence referred to a movement emanating from Auckland for the Tongariro River to be called the Upper Waikato.[5] The reports did not name any particular party, but noted the proposal emanated from Auckland and was supported by anglers from Whanganui. In view of later events, it is possible that Cecil Whitney, then deputy president of the Upper Waikato and Tongariro Anglers Club (UWTAC), may have been agitating for the change.

Johannes C. Andersen v James Cowan

Johannes C. Andersen publicly supported use of the name Waikato. He was the first librarian of the Alexander Turnbull Library and a member of the Honorary Geographic Board of New Zealand.[6] Of Danish birth, Andersen had taken an interest in many matters indigenous to New Zealand, from bird life to Māori songs and place names.

In newspaper correspondence, Andersen contended that 'Tongariro' was not the proper name of the river. That name could be ascribed to a 'comparatively insignificant river flowing from the Kaingaroa plains on the east, so that . . . there was already a river of that name'.[7] The Honorary Geographic Board requested him to form a subcommittee with Sir Frederick Chapman to look into the question of the river's name. Their research suggested that at all times in the past the name used for what by the 1920s was called by anglers the Tongariro was Waikato or Upper Waikato. Andersen buttressed his

Proponent of the Upper Waikato River name, Johannes Carl Andersen.
Alexander Turnbull Library, Wellington, New Zealand. /records/23130148

conclusion by viewing the Waikato River as a waterway starting as a tributary stream on Mt Ruapehu to the south of Tūrangi, and ending where the river joined the sea many kilometres distant. In effect, there was one long river which included New Zealand's largest lake:

> The European custom in naming rivers is to follow the river from its mouth to its source by way of the longest tributary, the one name being applied to that stream of water from the source to the mouth. The river now called Tongariro is the longest tributary of the Waikato, so although Lake Taupo intervenes it was logically called the Upper Waikato.[8]

Relying on this report, the Honorary Geographic Board retained the name Upper Waikato, but also had a bet each way by acknowledging that the river was also referred to as the Tongariro. On a government map printed in 1930,[9] the middle reaches were tagged as the Tongariro River, while the words Upper Waikato applied from above the junction with the Poutu River down to the lake. This, it seems, was a sop to the acknowledged popularity of the Tongariro as a world-famous angling river. However, the insertion of the words 'Upper Waikato' near the delta cannot have pleased the pro-Tongariro faction.

Andersen's claims were disputed with equal force by another man keenly interested in Māori history and customs, James C. Cowan. The author of many books and monographs on Māori matters, Cowan was familiar with the central North Island thermal regions.

Cowan acknowledged that both names had been used at different times by Europeans and by Māori. He, however, advocated the name Tongariro, because most of the water which passes from the river into the lake came from the Tongariro mountain range or its environs.[10] In his view, the Tongariro had its source with the Mangatoetoenui headstream on the upper slopes of Mt Ruapehu. Proceeding northwards, this waterway was joined by a side stream on the Rangipo plains, and it was this, asserted Cowan, that was the true Waikato River. Other streams rising from the bases of the mountains Ngauruhoe and Tongariro then joined, followed by a major tributary, the Poutu, from nearby Lake Rotoaira. This lake, Cowan pointed out, was fed largely by streams from Mt Tongariro, so in geographical terms Tongariro was the correct name for the river. His conversations with local Māori confirmed this.

The closest Cowan came to viewing the name Waikato as legitimate was his admission that it applied to one of the small streams forming part of the headwaters.[11]

A few days later, Andersen returned to the dispute. The Upper Waikato, he retorted, was the official name recorded separately by Dieffenbach and Wakefield, New Zealand Company officers who explored the thermal regions from as early as 1841.[12] J. Kerry-Nicholls, the explorer who investigated the district in 1883, also reported that the waters of the Upper Waikato came from the slopes of Mt Ruapehu. Andersen rejected Cowan's conclusions that the bulk of the water from the streams feeding the Tongariro came from the Tongariro Range, with the Mangatoetoenui being the largest tributary and the longest. Although he had never been to the locations concerned, the stream called the Waikato, the Mangatoetoenui 'and other considerable streams' had their source in the Ruapehu block. In effect, Cowan had wrongly downplayed the significance of the Waikato headstream rising in the Onetapu desert region in order to buttress his own theories favouring the Mangatoetoenui. Andersen also questioned why, if Tongariro was the true name applied by local Māori from the earliest times, it was not recorded as such by the first Pākehā surveyors who travelled through the district.

Andersen conceded that Māori had given the name Tongariro to a small part of the river from the mouth to a point 5 miles upstream.[13]

This section was the result of the Tongariro tributary rising in the Kaimanawa Range to the south-east of Tūrangi joining the Waikato Stream proceeding from the Ruapehu block to the south.

After two others had contributed their views on the subject, Andersen repeated his core argument. Tongariro may have been the name applied by some people to part of the river around the delta, and perhaps for some distance upstream, but the majority of the river's course from the source to that lower region was and had always been known as the Upper Waikato. Hochstetter and other explorers confirmed that usage. Nor did it make any difference that the stream called the Upper Waikato began as a tiny rill on the slopes of Mt Ruapehu, with larger volumes of water coming from tributaries such as the Waipākihi, rising in the Kaimanawa Range:

> the insignificance of the Waikato on Ruapehu as compared with the Waipakihi, . . . does not affect the nomenclature, as it is the main stream of the river, even if it is the smaller, rising in the main watershed, that is taken as the main river.[14]

Māori views

It is intriguing that local Māori opinion was equally divided over the naming issue. In early December 1931 a group representing Ngāti Tūwharetoa, the tribe which inhabited the Taupō district, petitioned the prime minister, the Hon. G. W. Forbes, rejecting any move to alter the name of the river from Tongariro to Waikato. The group's petition was organised by hereditary paramount chief Hoani Te Heu Heu, grandson of the Māori leader who had gifted the land for Tongariro National Park to the New Zealand government. [15]

The petition claimed that the name Tongariro River, as applied to the waters flowing from the Ngauruhoe/Tongariro/Ruapehu region 'has been associated from time immemorial with our people, their history, their legends, and their folklore, and we desire emphatically to protest against any change in a name that has, by long and endearing association as well as geographical fact, become so well established'.[16]

Hoani Te Heu Heu was the first name on the list of signatories. Others included Puataata Te Kerehi (P. A. Grace) and T. Hoka H. Downs; names which were reflected in the pools which anglers fished in the lower river below the main highway bridge.

The petition of Te Heu Heu provoked opposition. In January 1932, 26 Taupō Māori led by one Tupara Maniapoto presented their own petition, asking for a decision that the name 'Waikato' not be replaced by 'Tongariro'. The letter accompanying the petition made the telling point that Waikato had always been the correct name, with Tongariro being applied to a comparatively short stretch of water from the Tūrangi road bridge downstream to the land owned by the Grace family. It was this water which Robert Jones had leased and made available to anglers staying at Hut Camp or Tokaanu Lodge. These parties had a clear commercial interest in maintaining the Tongariro name which they had so successfully publicised.

Maniapoto's lobby group endorsed the view which Johannes Andersen had taken when the latter had first joined the debate:

> . . . there is a small clique at Tokaanu trying to get people to believe that an effort is being made to change the name of the river from Tongariro to Upper Waikato, but the boot is on the other foot. It is the clique which is trying to get the Waikato River called the Tongariro . . .[17]

Unworkable compromise

After the violence meted out to its signs in 1927, the Automobile Association seemed to take the hint. Some four years later, the Auckland branch council voted that on all official AA publications the river should be called the Upper Waikato from its source on Mt Ruapehu to its junction with its major tributary, the Poutu River. From that point down to the delta at Lake Taupō's southern shore, the name Tongariro would apply.[18] The name Waikato would also apply to the waterway formed by the outlet beginning at the northern end of the lake near Taupō town. In announcing this compromise, the council conceded it had noted the earlier decision of the Honorary Geographic Board of New Zealand on the issue. The association's touring manager, R. A. Champtaloup, erected signs giving effect to this decision at the Tūrangi road bridge.

True to form, C. A. Whitney tried to persuade the association's members at a meeting held in February 1932 that the proper name for the entire waterway was the Waikato River. But the publicity given by Zane Grey in his books and articles, and by other anglers, meant that Tongariro as a name had established a firm foothold internationally.

The Automobile Association felt there was too much to lose now that the river called the Tongariro was well-established as a world-renowned trout fishing venue.[19]

Officially then, the river was saddled with two names, neither clearly superior to the other, and each with its supporters. The Automobile Association, whose signs were unofficial but the only ones visible to the general public, called the river the Tongariro. This state of affairs persisted until late in 1944 when events took a new turn. Tūrangi residents and members of the anglers club found a new sign had mysteriously appeared on the Tūrangi main highway bridge bearing the name 'Waikato River'. Underneath in smaller letters were the words 'Tongariro Reach' followed by an explanatory statement which, as one angler noted, appeared to be designed to meet anticipated objections.[20]

The replacement signs were the work of C. A. Whitney. In 1944, after lobbying the Honorary Geographic Board yet again, he persuaded that body to alter the names so that the stretch from the delta to the Poutu River junction (the present Poutu Pool), formerly the Tongariro River, was described as the Tongariro Reach of the Waikato River.[21] This was a partial victory in favour of the name Waikato, which now assumed superior status. The Automobile Association, wishing to appear scrupulously impartial, altered its signs at considerable expense, even accommodating Whitney's request for urgent work to be done to allow the signs to be posted on the Tūrangi bridge before Christmas 1944.

The new signs brought a storm of protest to the AA from anglers and Tūrangi residents. The association's Auckland office, perhaps a little smugly, deflected the stream of angry correspondence to the Honorary Geographic Board — without revealing that Whitney was the cause of the problem — and awaited results.

Frank Yerex, Conservator of Wildlife, also now weighed in with his opinion. He favoured retention of the name Tongariro alone. He saw that name as deservedly world famous, and added there was no precedent for applying the same name to a river entering a lake and another draining it on the opposite side.[22]

This 1944 decision of the Honorary Geographic Board is an odd one. It seems to have been made after lobbying by C. A. Whitney, representing the pro-Waikato camp, and with little in the way of submissions from other parties. Possibly, the previous findings by

J. C. Andersen and Sir Frederick Chapman had already tilted the balance in favour of Waikato, but if that were the case it is difficult to see why it was not made in 1928, the time of the recess committee's original report. Alternatively, the board may have felt that, provided both names were being retained, it did no harm to simply clarify the extent to which the name Tongariro should be applied and where it stood in relation to the name Waikato.

Whatever the board's intention, it was not well received. The UWTAC members debated the issue at their Easter 1945 annual general meeting — one of the few times in the year when most members were grouped together. The gathering emphatically opposed the name Waikato. Several months later the club made a detailed submission, citing the undoubted notoriety which the Tongariro River had attained with overseas anglers, sporting writers and tackle manufacturers (such as the venerable English firm, the House of Hardy).[23] A key point made was that the New Zealand Government Tourist and Publicity Department had distinguished between the Waikato and Tongariro rivers in its publicity material.

The threat of ongoing confusion over the names drove the Honorary Geographic Board to review its earlier decision and achieve finality in the matter. In October 1945 it concluded that the lower river should be called the Tongariro from the junction with the Waihōhonu Stream down to the delta. The section above leading to the source on Mt Ruapehu would be known as the Waikato.

In reaching this split decision, the board cited one of the first European explorers in the Taupō region, John Carne Bidwill, who recorded that Māori living there used the name Waikato. But another European, Ferdinand von Hochstetter, stated that some Māori applied the name of Tongariro to a tributary rising in the Kaimanawa Range and also to the main river from the junction of the two waterways all the way down to the lake. The tributary is now known as the Mangamawhitiwhiti Stream.

The board also relied on Māori sources when reaching its decision. One of these was a song composed by a chieftainess called Ngawaero in which she referred to the part of the river near the delta as the Tongariro. The board explained:

> Although it has long been recognised that the Upper Waikato was a Geographical Part of the lower Waikato River, yet early Maori usage and

> later some European usage had called part of the Upper Waikato the 'Tongariro'. Thus we have the position of a river being called 'Waikato' for a distance of 35 miles downstream from its source, then changing its name to 'Tongariro' and continuing as such to Lake Taupo. It then flows from the lake as the 'Waikato'.
> . . . fishing licenses and notices in Official Gazettes have established the usage of Tongariro for portions of the Upper Waikato . . .
> The Board therefore in making its decision gave due regard to the confusion that would continue to arise as these fishing areas were extended and it therefore decided that the name 'Tongariro' River should apply to the river from the delta at Lake Taupo up to the junction of the Waihohonu Stream rising on the slopes of Tongariro Mountain and the Waikato River. The Waikato River would still retain its name from its source down to this junction.[24]

So at first the Honorary Geographic Board of New Zealand decided to name the river the Upper Waikato (Tongariro), in 1928. It then revisited the issue in 1944 because of possible confusion over which waterway near Tūrangi was the Tongariro for angling purposes (and possibly because of Cecil Whitney's renewed submissions). In October 1945 the board tried to settle matters once and for all, applying the name Tongariro to the lower section from the delta to the junction with the Waihōhonu tributary. The name Waikato continued to apply from the source to the tributary junction. It was an uneasy compromise, with the historical records of early usage yielding to the notoriety of the more accessible lower reaches used for trout fishing.

While the controversy over the proper name for the Tongariro River rumbled on, trout continued to run the river in autumn and people carried on fishing for them. The size of trout available was underlined in 1932 when, after a small flood, Beauchamp Urquhart Barlow found a large brown trout dead on the bank. It was a superb specimen, 25lb in weight, 33½ inches long and 22 inches in girth. These dimensions gave it an exceptional condition factor of 67.

Barlow sent the fish to Conservator of Wildlife, Andy Kean, who put it on public display. Kean said it was the heaviest brown trout authentically recorded in the Rotorua district in the previous 25 years.[25]

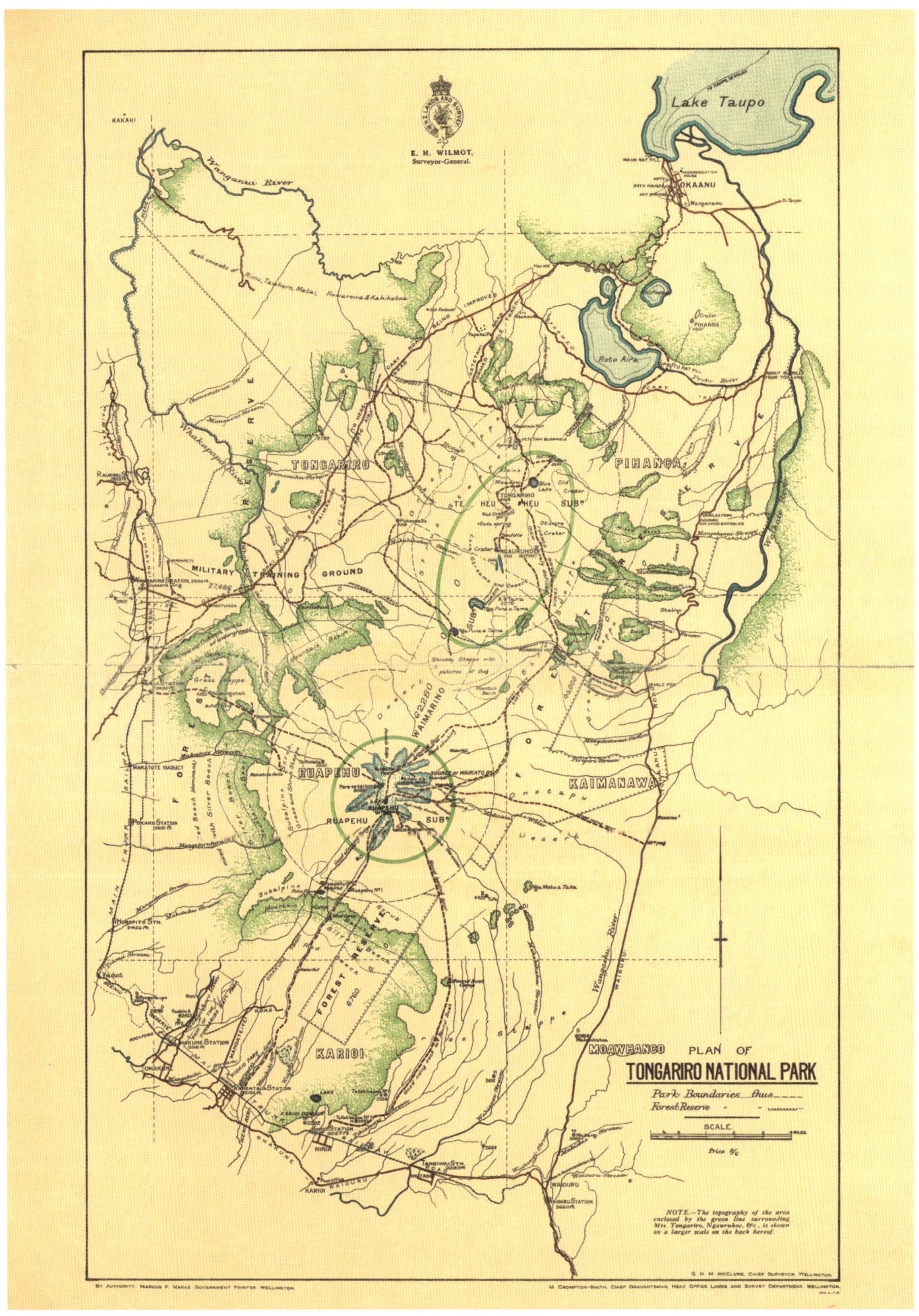

Plan of Tongariro National Park, New Zealand. The path taken by the Tongariro River can be seen at upper right of the image. The hairpin bend (referred to in chapter 1) taken by the Waipākihi river, flowing from the Kaimanawa Ranges, appears at centre right.

Department of Lands and Survey cartographer, Series NZMS 54, 1917. Image sourced from LINZ

The Dreadnought Pool, January 1938. Joyce Galbraith collection

All that remains of the 'Dreadnought' thirteen years later, April 1951. Joyce Galbraith collection

Joe Frost at his Turangi tackle store in 1939, with abundant evidence that the dry fly would take trout on the Tongariro River.

Whites Aviation Collection, Alexander Turnbull Library, Wellington, New Zealand, Reference Number WA-05235

In the 1958 flood the Bridge Lodge came close to being swept away.

Joyce Galbraith collection

Joe Frost at his Turangi shop, 31 October 1947, watching one of his three staff tying a fly. Newly completed lures rest on a rack in the background.

Whites Aviation Collection, Alexander Turnbull Library, Wellington, New Zealand, Reference Number: WA-10434

Looking south on Taupahi Road from the Tongariro River bridge, 12 October 1948.

Wellington-Taihape-Rotorua-Cen: Pihanga from Tongariro bridge, photo by Leslie Adkin. Museum of New Zealand Registration Number A.007390

Tongariro River anglers with their end-of-season catch, May 1941. The average weight of the trout was 7½lb.

W.B. Beattie, Auckland Libraries Heritage Collections, ref: 1370-626-03

Peter McIntyre failed to appreciate Tom Shand's artistic efforts.

Cartoon by Neville Lodge, The Evening Post, 8 July 1964.

White water at the head of the Hut Pool, Tongariro River, Waikato. Photographer unknown: Views of Tongariro River, Waikato.

Alexander Turnbull Library, Wellington, New Zealand, Reference number 1/4-018069-G

The 1958 flood damaged both river and houses.

W. G. Henderson

The Upper Waikato Stream near its source on the North Island's volcanic central plateau. The stream rises on the eastern slopes of Mt Ruapehu, seen in the background wreathed in cloud.

Estate of John Parsons

The Upper Waikato Stream where it crosses State Highway 1, to the south of Tokaanu. This location is considered to be the site of the first rainbow trout liberation by Malcolm and Forrestina Ross in the Tongariro watershed in February 1898.

Estate of John Parsons

A Hardy Phantom minnow bristling with treble hooks.

Thomas Turner Fishing Antiques, England

The Silex No 2 spinning reel (circa 1913), made by the House of Hardy and used by many Tongariro anglers.

Thomas Turner Fishing Antiques, England

Survey map showing the lower Tongariro River from the main highway bridge to the delta, 1920.

Puketi Survey District, Hautu Blocks, Maori Land Plan ML3438, code R22521992, 18 March 1920, Archives New Zealand

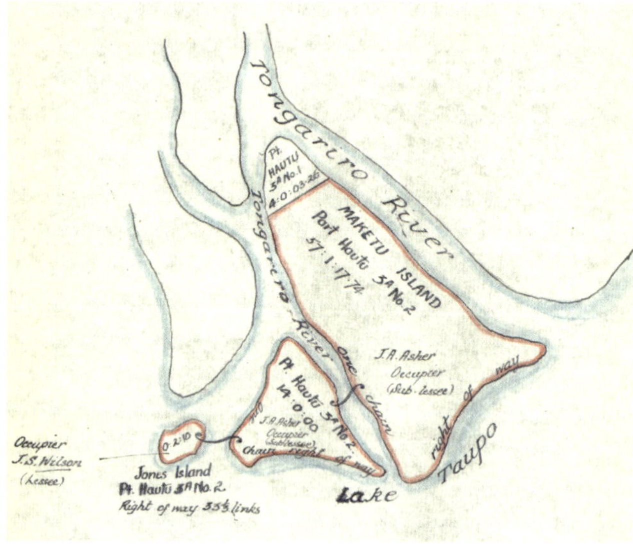

The Delta Camp was located on the smallest island at the mouth of the Tongariro River. Called Jones Island, it measured 2,276 square metres.

Lease to Hautu 3A2 – J A Asher, code R21530170, Archives New Zealand

Showing fishermen trout fishing in Dreadnought Pool on the Tongariro River near Tūrangi.

Auckland Libraries Heritage Collections 996-208

The Bridge Fishing Lodge on the true left bank came close to being swept away in the flood of 1958.

W. G. Henderson

Scientist Michel Dedual inserting a radio tag in a Tongariro River rainbow trout.
Department of Conservation

The Tongariro River in heavy flood (looking upstream from Red Hut bridge) during the 1980s.
Estate of John Parsons

Red Setter and Glo Bugs.
W. G. Henderson

Footbridge above the Major Jones Pool.
Auckland Libraries Heritage Collections 1021-901, Eric W Young, 1980s

Impoundment on the Waihukahuka Stream, Tongariro National Trout Centre, June 2020.
W. G. Henderson

The angler's wish: a Taupō rainbow in excellent condition.
W. G. Henderson

Fishing in sunshine above the Red Hut Pool.
Estate of John Parsons

Paintings of the Tongariro River by artist Val Raymond, which she gifted to the Tongariro National Trout Centre.

Courtesy of the Val Raymond family.

CHAPTER 11:

Neilson, Begg and Hickling

'. . . out would go their lines, straight and sure and true, a hundred feet into the deep and fast water, their flies dropping like thistledown within a foot of the far bank.'

— HAROLD HICKLING, *FRESHWATER ADMIRAL — FISHING THE TONGARIRO RIVER AND LAKE TAUPO*

SEVERAL ANGLERS BECAME FIXTURES on the Tongariro River in the years before and after World War II. James Neilson and Robert Begg were two of them. Harold Hickling came later, helping to boost the fame of the fishery by writing about it in the 1960s.

James Neilson

James Swan Wilson Neilson was living proof of the adage 'Allah does not deduct, from the allotted span of Man, those hours spent fishing'. He first fished the Tongariro River in 1910, continuing until his death in 1957, making him one of its most devoted adherents, and one of very few anglers who would have observed the fishery in its two prime periods: pre-World War I and in the early 1920s. On one visit to Tokaanu in May 1923 he averaged 10 fish each day.[1]

Neilson was born in Scotland on 7 November 1869, and farmed there before emigrating to New Zealand in 1887. In 1905, in partnership with another, he started Tawaroa Station, a 6,000-acre block in the Bay of Plenty stocking sheep and cattle.[2] A near-fatal car crash in 1938 forced him to give up farm management, but must have left more time for fishing.

On three occasions he won the Whitney Tongariro River Cup for the heaviest rainbow caught below the Birches Pool. These catches were made in the 1937–38 season (10lb 2oz from the Log Pool), 1938–39 (9lb 10oz, Judges Pool) and 1941–42 (10lb, Hut Pool).

Neilson would book in at Tūrangi in time for the opening day, fish all through the season, and then book out on the closing day. Always the same room at the lodge, always the same place at the dining room table.[3] Alpha Cuthbertson, whose husband owned Taylor's Lodge, would usually drive him up to Rotorua, where he would spend the closed season. Neilson held the firm view that no one could be called an angler until he or she could rise and land a trout on a barbless hook.[4]

In 1950, Archie Kitto wrote to the Minister of Internal Affairs, the Hon. W. A. Bodkin, asking for the Island Pool to be renamed using Neilson's name. The minister had no power to effect such a change and referred Kitto to the official body dealing with New Zealand place names, the New Zealand Geographic Board. Nothing came of this, but it shows the high regard in which Neilson was held by other anglers. Harold Hickling referred to him as the doyen of Tongariro anglers.

Neilson died in 1957, unmarried, and was buried at Rotorua.

Robert Leith Begg

Robert Leith Begg took out his first fishing licence in 1897 at the age of 17. Born in Roxborough, he was one of 12 children born to a tea merchant. Trout fishing was not his sole interest, for while in Southland he became a racehorse owner with some success.

Begg was educated in the Canterbury town of Ashburton and went on to lead a varied business life. He began as an auctioneer in Gore, but in the late 1920s he farmed a property at Maxwell, near Whanganui, before moving to Wellington several years later.

Proximity to the fabled fisheries of the Taupō district meant that this became his favourite fishing haunt, with the Tongariro accorded special attention. His angling skills, carefully crafted and tested on the brown trout of Southland, easily transferred to the northern rainbow waters, a fact borne out years later when, in 1949, he reportedly caught his 10,000th trout from the river.[5]

A guide for US anglers

In 1923, Begg caught a 20lb rainbow trout from his beloved Tongariro — surely his largest. This specimen subsequently went on display in the Whanganui museum. His angling prowess reached the ears of officialdom, and on several occasions he acted as angling adviser to Governors-General Lord Bledisloe and Lord Galway on their visits to the Taupō–Tongariro area. Of course, local knowledge is vital in angling, and if speedy results were wanted at short notice, the officials at the Department of Internal Affairs could not have picked on a better man to guide the VIPs on their fleeting visits.

Fortunately, there remains a motion picture record of the exploits of this diminutive angler, who, despite his size, was able to cast a fly across the widest of the Tongariro's pools — no small feat in the days before the water flow in the river was halved by power generation requirements. Contemporary photographs show him in the river, a double-handed rod in hand and a heavy reel mounted on it. In 1949 the United States fly fisher Selwyn Kip Farrington and his wife, Chisie, came to New Zealand to fish for Tongariro trout. The couple stayed at Taylor's Lodge. At the request of the government Begg acted as a fishing guide, and appears casting a line on the Tongariro in the Farringtons' short film *New Zealand Rainbow*.[6]

Upper Waikato and Tongariro Anglers Club

Begg was a member of the Upper Waikato and Tongariro Anglers Club, the executive officers of which were elected by postal ballot notified in the *New Zealand Fishing & Shooting Gazette*.[7] Begg was on the Wellington section of the executive, along with F. E. Thornton, the magazine's editor. They were charged with interviewing the Department of Internal Affairs on matters raised by the club executive and reporting back.

At the inaugural AGM of the club on 11 May 1933, Begg delivered a paper on 'The Etiquette of the River', directed to pointing out the courtesies required of all anglers.[8]

Robert Begg (at left) receiving a trophy for the largest trout of the season.
NZME copyright, (NZH, 4 Nov 1933)

Robert Begg at work on the Tongariro River.
Auckland Libraries Heritage Collections 1370-626-10

At a special meeting on 4 November 1933 Begg was presented with the Whitney Upper Waikato Cup for catching the largest rainbow trout from the Upper Waikato in 1932–33. This trophy was donated by Sir Cecil Whitney of Auckland for the angler who caught the largest rainbow above the Birches Pool. Begg caught his prize-winning fish on artificial fly on 22 May 1933 in the Dreadnought Pool — weight 7lb 13oz.[9] The following season he collected the Whitney Tongariro River Cup, for the member who caught the heaviest rainbow below Birches Pool. Begg's winning fish was 11lb 12oz.[10]

Along with several other anglers (H. J. Duigan and Rangi Arapeta of Tokaanu), Begg discovered the pool that marked the upper limit of fishing on the river and which now bears his name — Beggs Pool (referred to as Begg's Pool in some publications).[11] The location was above the Mangatawai Stream and below the Waihua Stream (a Tongariro tributary).[12] Robert Begg died aged 73 years.

Harold Hickling

Harold Hickling was a product of the Victorian Empire and clearly proud of his place in it. Born in 1892, he attended the naval college Osborne on the Isle of Wight, and then the naval college at Dartmouth, entering the Royal Navy as a midshipman in 1910. He was of an age to see service in the Royal Navy in both World Wars.[13]

War service

One of Hickling's major achievements in World War II was overseeing the installation of the Mulberry artificial harbour shortly after D Day in June 1944. This project was vital for the successful supply of Allied forces in Normandy, and survived a huge Atlantic storm two weeks after the landings had been made.

Hickling left the Royal Navy in 1948 with the rank of Vice-Admiral (Rtd), and arrived in New Zealand on a fishing holiday in 1950.[14] Although he had married a South African lady, Gwynneth Tennant, in 1920, it seems he left her in England and does not seem to have divorced her at the war's end. That may explain why he had a de facto relationship with Mary Anstruther.

New Zealand residence

At first Hickling had intended only a three-month holiday, but decided to reside permanently in Tūrangi, then a small trout fishing village. The principal function of the town in those days was to service the needs of the increasing numbers of tourist anglers, both New Zealanders and those from overseas.

Hickling purchased a house hard by the Major Jones pool, already renowned as a prime fishing location, and soon learnt the gentle art of swinging a wet fly through a likely lie. He was no stranger to the art of angling, having learnt to fish in England and enjoyed the sport in a number of countries. However, this experience was of little use to him when he found himself confronted by the fiercely flowing Tongariro and large trout whose capture required fishing tackle akin to that used for salmon.

A few of the locals offered assistance and encouragement, but Hickling appears to have learnt by observing others and applying their methods until success finally arrived. One of his Tongariro teachers

Mary Anstruther at right, admiring a Tongariro rainbow.
Estate of John Parsons

was an experienced angler and farmer from Whanganui, Archie Kitto. After Kitto had hooked a fine trout in the Island Pool he let Hickling take his rod and land it. The fish was later produced to guests at a riverside lodge as one Hickling had caught himself. He admitted the catch meant he was sailing 'under somewhat false colours',[15] but went on to catch many trout through his own efforts.

After settling into Tūrangi, Hickling met Mary Anstruther at a casting tournament, and the pair decided to live together. Mary, born in Yorkshire in 1902, had arrived in New Zealand after World War I and later married, but had been widowed at an early age. She, too, was a keen trout angler and would often join Hickling on the river.

Writing in retirement

Harold Hickling the angler is chiefly remembered for his book *Freshwater Admiral — Fishing the Tongariro River and Lake Taupo*, published in 1960. This was the product of 10 years' knowledge accumulated by intensive trout fishing in one location. O. S. Hintz, newspaper editor and author of the classic *Trout at Taupo*, had encouraged him to write a book on the subject.

Mrs Anstruther and Vice-Admiral Hickling fishing in the Tongariro River.

Photographer T. Ransfield. Archway Item ID:R24459275 Archway Series Number:6539

If his intention was to provide a practical guide to fishing the river, then the project was doomed from the start. Beginning in early December 1957, Hickling took three months to complete a manuscript for publishers A. H. & A. W. Reed (initial titles were 'The Tongariro Tale', later changed to 'Fishing the Tongariro'). The fates then intervened to torpedo this first effort. A massive one in 500-year flood on 24 February 1958 completely altered the Tongariro riverbed, destroying established pools and making the text irrelevant except as a piece of angling history.

Undeterred, Hickling and others set about surveying the river in its new bed, assessing the pools for trout holding places, and generally gathering an overall picture which would form the basis of a revised manuscript. The resulting work presents an enchanting mixture of fishing instruction and anecdote. The main focus is on the techniques required to fish successfully the downstream wet fly on the Tongariro River, but it also contains discussion of other fishing localities on Lake Taupō's southern shore, as well as giving guidance (now mostly outdated) to people intending to visit New Zealand.

As a permanent resident of the Tūrangi angling fraternity, 'the Admiral' was often sought out by novice and experienced anglers alike for his angling knowledge and advice. It was not always a smooth exchange. For example, his expertise in this regard, not to say his diplomatic skills and linguistic abilities, were tested to the extreme when attempting to guide a visiting Frenchman and his wife through their first encounter with the complexities of Tongariro angling.

One of his characteristics was the use of the left hand to cast the fly rod and the right hand to wind the fly reel. To any left-hander, such as the current author, this is a perfectly natural thing to do, but the apparent ease which the method accords is not available to right-handed anglers. Taupō author and angler John Parsons discussed the question of right- and left-handed casting in correspondence with the Admiral. Only when they discussed the matter in depth did it emerge that Hickling had been unconsciously applying the method which Parsons had been suggesting.[16]

Harold Hickling belonged to the era of long cane rods and heavy fly reels, when tackle makers like Hardy and Farlow were the automatic choice of anglers, and the deep-sunk wet fly was the primary angling technique (noted visitors such as G. B. Hobbs aside).[17] The value of Hickling's book lies in its careful recording of angling methods and personalities in a time before the flow of the Tongariro River was altered forever by the demands of hydro-electric power generation. As an active member of TALTAC (Tongariro and Lake Taupo Anglers Club) he did much for the interests of the Tongariro fishery.

His son William Tennant Hickling, only 18 years old, died along with almost 900 others when the battleship HMS *Barham* went down in the naval action off Sollum in December 1941.[18] William was apparently named after a naval friend and superior, Admiral Sir William Tennant. With Britain's naval position in the Mediterranean so finely balanced at the time, no official disclosure of the ship's loss was made; something which made Hickling's loss all the more difficult to bear.

Harold Hickling died in his home at Tūrangi on 12 November 1969, aged 77 years, not far from the river which had yielded him so much beloved sport with the fly rod. He lies buried in the Taupō cemetery. Mary Anstruther survived him and remained in Tūrangi until her death in 1994.

CHAPTER 12:

The Upper Waikato and Tongariro Anglers Club

Anglers fishing Taupo and its rivers have formed an Anglers' Club, in an endeavour to see what can be done to improve existing conditions. . . . it is hoped that all interested will join.

— ***THE AUCKLAND STAR*, 6 JUNE 1932**

THE PUBLICITY FROM ZANE GREY'S FIRST New Zealand visit generated increased interest in the fishing opportunities of the Taupō region, and the Tongariro River in particular. The first motor car had appeared in Tokaanu in February 1904, but by the 1920s New Zealanders had adopted this from of transport with gusto, with vehicles becoming much more common. Although the unsealed roads on the central plateau could still make for a difficult journey, especially after rain, motorised transport meant that groups of anglers were able to fish the Taupō region more frequently.

In 1907 Tokaanu was a European hamlet that was home to a mere 44 people.[1] It was one of several settlements on Lake Taupō's southern shore, almost all of them Māori hapū, each with its own marae and houses.[2] The village of Tūrangi, known locally as Taupahi, also began to grow in the late 1920s and early 1930s after the Crown auctioned land there.

As angler numbers increased, so too did the need for local accommodation. There were three types: fishing lodges, private cottages and informal lakeside camp grounds.

Fishing lodges had long been a permanent feature in Tokaanu, although the limited fishing season meant they had patronage from anglers for only part of the year. By 1931 Tūrangi was gradually changing into an angling settlement, with summer cottages offering accommodation of a kind much better than the rudimentary anglers' tents and cabins used in the early days.

The lodges, too, began to improve their facilities. The Delta Camp business carried on after Robert Jones's death in 1924, managed by local identity John Asher. By 1929 he had competition from the Tongariro River Lodge (the old Hatch's Camp), which was located above the main highway bridge near the Island Pool, and later became Taylor's Lodge, home to many visiting anglers. As soon as James W. K. Taylor bought the business in 1931, he advertised a major improvement: hot and cold showers for guests.[3]

Fishing club formed

An advertisement in the *New Zealand Fishing & Shooting Gazette* of May 1932 sought interest from persons who might wish to join a proposed Upper Waikato Anglers Club. The Upper Waikato and Tongariro Anglers Club (UWTAC), formed in time for the start of the 1932–33 fishing season, began life as an unincorporated association of fly fishers who were drawn each year to Tokaanu to fish the great river. Some 100 members joined initially; most of them New Zealanders, but some were overseas anglers who made annual visits to the Tongariro River.

Cecil A. Whitney was a founding member who presented two silver cups as trophies for competition angling by club members. In line with his views on the proper name of the Tongariro River, one cup was for the heaviest trout taken in the Upper Waikato River (from the source to the junction with the hatchery stream), while the other was for the heaviest caught in the Tongariro.[4] In early June 1933, the Waipawa angler J. T. Pickett received the Tongariro cup for banking a fish of 11¼lb from the Fence Pool. The Upper Waikato cup was won by the Wellington expert Robert Leith Begg, with a trout weighing 8lb taken from the Cliff Pool. Both of these trophies are still keenly contested by the club's members.

New Zealand's status as a Commonwealth Dominion meant that among the club's overseas members those from the United Kingdom featured prominently. In F. E. Thornton, a Wellington schoolteacher, the club had a capable secretary. As a teacher and editor of the *New Zealand Fishing & Shooting Gazette*, Thornton could correspond with anglers on points they raised about fishing, while also writing to civil servants about issues affecting the Taupō fishery.

As head of the Colonial Ammunition Company (CAC), Cecil Whitney was another founding member whose links would prove valuable. The CAC had been engaged on munition supply contracts with the New Zealand government during World War I, which meant Whitney was well-heeled and could also provide the political connections that could prove useful to the club. Together, these two men would come to provide a constant chorus of opinion on the state of the Tongariro fishery and how the Department of Internal Affairs should manage it.

One staff member of the Wildlife Branch (of the Department of Internal Affairs) was not enthusiastic about the club, seeing it as a potential source of trouble. The Conservator of Fish and Game, Andy Kean, wrote to a superior:[5]

> I anticipated trouble with these people the instant the formation of this Club was proposed. The majority of the prominent members are persons who have been connected with various acclimatisation societies and have shown very little ability in administrating the affairs of their own District.
>
> . . .
>
> Personally, I very strongly object to the action of this Club in interfering with our District and feel that unless the Club is made to recognise the fact that we are perfectly capable of managing our own affairs they will gradually obtain an undesirable footing which will, in time, be most detrimental to this Department.

The UWTAC held its first annual general meeting at Taylor's Lodge, Tūrangi, on 11 May 1933. Of the more than 100 members, a mere 15 attended. Of those present, three men — Robert Leith Begg, Cecil Whitney and F. E. Thornton — ensured the introduction of a predetermined agenda, a rather contrived action which seemed obvious to another member.[6] However, things did not go quite so smoothly as the triumverate had planned. When Whitney opened proceedings by proposing certain rules to limit the number of anglers who could fish each pool in the river, B. U. Barlow reminded the meeting of the practical limitations of this approach. The club, he said, had no jurisdiction over the fishery, and any motion passed by the meeting could not be more than a suggestion which the Department of Internal Affairs could adopt or ignore as it wished.

The meeting went on to discuss a proposal that certain rules of angling etiquette propounded by Whitney and the others should be printed on the back of anglers' fishing licences. The nature of the rules, apparently drafted by Robert Begg,[7] is a clear response to the issue of overcrowding on the river; an issue that became more acute as the Taupō fishery developed and became ever more popular.

Draft rules for Tongariro fishing

A proposal to limit the hours of fishing to a period from 4am to 9pm was carried only when the chairman used his casting vote. Other matters of concern to the meeting were the use of the ranger on anti-poaching duties over the winter (closed) season, rather than being seconded to work on the trout hatchery and measures to increase the amount of trout food available.

The draft code of angling etiquette decreed:

> We urge all fishermen to remember there are other rods on the River, and to at all times please remember fishing etiquette.
> The Club asks everyone to be particular in observing the following Recommendations:—
>
> All incoming anglers consult the man already in the pool as to where he should start.
> No incoming angler to get ahead of an occupant already in a pool without his permission.
> No angler to occupy a pool indefinitely, but to move on after every cast. By this means every man on the pool gets his chance at all the good portions of the pool.
> Time limit for a party for a pool is half a day.
> First angler to reach a pool and take up a position showing that he proposes to fish there has first choice of positions.
> When the pool is filled to rod capacity no newcomer may enter it without permission.
>
> We would suggest the full capacity for each pool as follows:—

Major [Jones]	3 rods	Log	4 rods	De [Latours]	2 rods
Island Pool	2 rods	Stump	1 rod	Downs	4 rods
Judges	3 rods	Crescent	3 rods	Graces	3 rods
Weir	2 rods	Reed	2 rods	Willow Reach	3 rods
Bridge	2 rods	Gun	2 rods	The Bend	4 rods
Swirl	1 rod	Jones	3 rods	Gull Beach	3 rods
Hut	4 rods	Nellies	1 rod	Poplar	4 rods
Boat	2 rods	Fence	2 rods	Cherry	3 rods

This suggested allocation of fishing water allowed for 63 rods on the then available river.

Thornton met with Bennett, a Department of Internal Affairs official, and pressed him for consent to have printed copies of the 'rules' printed on the reverse of fishing licences issued to Taupō district anglers. Bennett thought the draft rules bordered on the ridiculous[8] and refused, so Thornton requested that they be interleaved with the licence books. The department was keen to keep well away from any possibility of controversy that the application of the 'rules' might bring. The best position Thornton could obtain was the neutral suggestion that he make a direct approach to tackle stores and other agencies selling licences for their co-operation in distributing the rules via those outlets.

The Department of Internal Affairs was anxious to avoid giving any official backing to the proposals, which were clearly seen as the potential source of countless riverbank squabbles. For the department, the role of referee in such encounters held no appeal.

UWTAC political initiatives

By the early 1930s a number of anglers were starting to feel that the Tongariro was no longer the river it once was. Catches, it seemed, were down both in the numbers and in the average size of trout taken. This view took hold among a substantial minority of anglers, a number of whom had experienced the magnificent angling of the mid-1920s and could feel a lighter weight in their creels. After the end of the 1934 season, tipped by many as the worst ever, anonymous letters began to appear in daily newspapers complaining of too many rods on the river, access difficulties, the disappearance of famous pools, and inappropriate fish stripping and stocking practices by the fisheries officers.[9] The general theme was that the Department of Internal Affairs was doing nothing to halt the decline in the quality of the fishing.

In an article published in a national newspaper, an anonymous angler caustically commented:[10]

> It is hard to believe that anybody owning the most famous rainbow trout river in the world, the Tongariro, would allow it to lose its reputation. Is there no one in authority who has sufficient interest in this great asset to see that the drift that has taken place during the last few years is arrested?
>
> The fees collected by the Government for angling in the Tongariro River amount to something in the vicinity of £1500 per annum, yet the Department has not spent one pound in improving the conditions during the last three years. At all times complaints can be heard from both local and overseas anglers about the shocking conditions that exist on this world-famous river, yet nothing is done.

The Auckland Star fishing correspondent reported:

> What probably has been the worst fishing season on the Tongariro and definitely the worst month of May, has finished. At long last anglers and visitors have realised the state of drift and neglect this most valuable asset to the Dominion has reached. Our overseas visitors are wondering if it is going to be worth their time to return and what they can conscientiously say to enquiring friends at Home.
>
> . . .
>
> Do the powers that be realise what they are throwing away from a tourist point of view? Next season will see visitors and sportsmen from all over the world coming to us after their visit to Australia, and what are we going to give them? An overcrowded Delta, a river empty of fish from a fishing point of view and congested with rods until the pleasure of fishing is gone; the roads full of holes, bad crossings with no facilities for any but the young and strong to cross them; and, worst of all, the fact that there is miles of fishing that the Internal Affairs Department will not open or allow anyone else to bring into use.
>
> . . .
>
> Is it not time something was done by the public to help the Tongariro Anglers Club and others to bring pressure to bear . . . to save this heritage for its owners' good?[11]

Some of the complaints were fanciful and almost speculative, such as the allegation that trapping trout and stripping ova from them at the Tongariro hatchery had an adverse effect on the fishery. Another was that with angling permitted for 20 hours out of every 24, this was too long a period as the fish needed more rest.[12]

Yet there could be no denying anglers' comparisons of their situation with the halcyon days of the mid-1920s. In the 1933 season Taylor's camp allegedly saw 4,200 trout smoked, while in 1934 the total was said to be less than 700.[13] To the anglers, at least, the dwindling size of their catches was inescapable.

River crowded

With regard to congestion of rods, it is clear that the water below the Tūrangi settlement was the scene of most anglers' efforts. The UWTAC's proposal to limit the number of rods in each pool started at the Major Jones pool, not far above the main highway road bridge. While upper pools like the famous Dreadnought were known to anglers, riverside trees and scrub made access and casting difficult. For this reason, pools like the Log and the Hut received the most pressure. Little fishing was done above the Tūrangi–Taupō road bridge.

As the Tongariro became a fishing Mecca, anglers started to complain that the river was too popular and subject to overcrowding, especially at holiday times like Easter. At the end of the 1928 season the *Fishing Gazette*, a British publication, charged the New Zealand tourist authorities with offering misleading information about Tongariro trout by basing the potential catches on records set decades earlier.[14] Officials responded by quoting a New Zealand angler who had taken well-conditioned fish in the first months of the season, and considered the quality of angling available to be as good as ever.

Others were not convinced, and viewed Taupō as a place where there was no room to swing a rod. That may have been true of favoured pools like the Log or the Hut, where people waited their turn to fish, but anyone willing to explore new and untried water was often rewarded.

Cecil Whitney did this by going up above the Dreadnought Pool in November 1932 with a group of friends. They were the first to fish the pool, later called Beggs Pool, below the falls, which were about 10 feet high and impassable to trout.[15] Graham Findlay of Miranda caught the first trout there and lost heavier fish in the process. The party had to travel south on the road to Waiouru before cutting east and walking down a track for an hour, which took them to the falls. That was their only access route; wading upstream from the Poutu Pool, it was impossible to gain access to Beggs Pool because the river passed through a series of gorges.

The UWTAC pressed the Department of Internal Affairs through meetings with the responsible minister, the Hon. William Parry. Club remits passed for the improvement of the fishery were regularly brought to the minister's attention by deputations of anglers appointed by the club for the task. A large number of these remits sprang from the mind of Lieutenant-Colonel Allan Oldfield, a retired British army officer described in one meeting with the minister as an old penfriend of the department. Shortly before the deputation met the minister in May 1936, Lieutenant-Colonel Oldfield died. At the same time, perhaps fortunately for the authorities, a large number of remits under his care disappeared with him.[16]

In a prompt response to the anglers' delegation, the minister dispatched his under-secretary, Joseph Heenan, to Tūrangi to investigate matters. Heenan met with anglers over several days and attended the UWTAC annual meeting on 16 May 1936. Heenan was a career public servant who had been transferred back to the Department of Internal Affairs in April 1935 after 15 years in another government department. This extremely capable public administrator was interested in a number of sports such as rugby football, but was no fisherman, his chief sporting interest being horse racing.[17] This, however, did not prevent him from forming his own opinions through careful observation and questioning of anglers. Staying at Taylor's Lodge, he offered a ready ear to all who cared to talk:

> I had previously heard a good deal on the question of congestion of anglers on the Tongariro river and some remarks on the matter were passed during my short stay . . .
>
> Others laughed at the suggestion. However, I was able to form a very good idea from my own observations as to what is involved. We motored down to Grace's Pool and there were not more than 5 anglers fishing along the long stretches of water from the bridge down to that point. From there downwards, there were only 2 anglers, and at the delta there were three launches. In the afternoon stretches of the river above the [Tūrangi–Taupō] bridge were examined, and apart from the

> Duchess Pool, Kowhai Flat, where there were three or four anglers, long stretches of the river were deserted. To complete the irony, on returning to Taylor's in the evening, I learnt that a Captain Godfrey had caught 11 nice fish down the river — he had had a wonderful day and had had the Gull Reach to himself. Another angler who had only caught two fish all day assigned the reason to 'the same old story; too many rods on the river.'[18]

Heenan's report concluded that all of the evidence pointed to Tongariro trout being in fine condition, with some anglers holding the opinion that the fishing was the best it had been for five years.

Nevertheless these complaints, even if largely unfounded, pointed to the need for a thorough assessment of the state of the fishery by fisheries scientists. Some of that work had already begun, with a comparison of scale readings between trout caught in the 1927–28 and 1930–31 seasons indicating that the average age of the fish had decreased from 4.2 years to 3.5 years.[19] But a broader, more comprehensive approach was several decades away. The probable cause of the poor results in the 1934 season was a prolonged period of dry weather, which caused the river to run at a low level for months. Expert anglers such as Robert Begg took the conditions as they found them and were rewarded for their persistence. Fishing improved near the season's close, with favoured pools, such as the Log and the Hut producing double-figure trout in May.[20]

Constant and sometimes contradictory complaints by Tongariro anglers sparked this cartoon from A. S. Paterson.
The Dominion, 21 May 1936.

The UWTAC's activities carried on in the interwar years. While fishing was of course the principal activity, the committee tried to improve things for its members in other ways, too. In late 1933 the president, Lionel Hanlon, asked the New Zealand Automobile Association to join the UWTAC in lobbying for improvements to the old coach road that ran from Waiouru to Tokaanu. Traffic on this unmetalled route had to contend with pumice soil, which became impassable after one day's rain and added to the time it took for angling parties to reach their destination. Not only would a better road cut the length of the journey, it would also help open up access to pools on the upper river near Beggs Pool. The government responded to the club's plea several years later by allocating funds for the roading improvements.

At the end of the 1934–35 season, Governor-General Lord Galway was elected patron of the club. A bag of 13 fine trout he and Lady Galway took from the Log Pool may have influenced his lordship's decision to accept the post.

To confound the pessimists in the fishing fraternity, the 1937 season opened with some good portents. A week after opening day, Frank Dyer, the UWTAC president, noted that trout were half a pound heavier than those taken at the start of the previous season, and some fly fishers had taken their limits.[21]

Effects of flooding

The Tongariro regularly experiences spate conditions, and there are severe floods on occasion. As the riverbed altered its course after each new deluge, anglers became concerned that popular pools in the lower reaches, like the Log and the Gun, would be lost. A large flood occurred in 1932, and another followed in March 1933, although in this instance the press reported some improvements had appeared in the form of an enlargement of the Duchess Pool. As the river changed, the UWTAC lobbied the government to protect famous pools like the Hut and Log — see Chapter 15.

Legal wrangles

As we saw in an earlier chapter, the Honorary Geographic Board of New Zealand had settled the controversy over the correct name for the river, Upper Waikato or Tongariro, in January 1946, so part of the waterway would be named Tongariro. This decision had repercussions for the name of the angling club.

At the annual general meeting held in April 1943, Grant Fargie took over as president from Mr Newton. The two vice-presidents were the elderly J. S. W. ('Jimmy') Neilson and local fly tier Joe Frost; C. A. (Cecil) Whitney was voted the curious post of deputy chairman.

Fargie was re-elected as president on 8 April 1944. At the annual general meeting held on that date, the members present voted in favour of a motion to dissolve the position of deputy chairman, at the same time voting thanks to Whitney for his past services and generosity to the club. Significantly, the meeting also voted to alter the name Upper Waikato and Tongariro Anglers Club to Tongariro Anglers Club.

Shortly after, it became apparent that the motions passed had not followed the proper procedures prescribed by the club's rules, having not been circulated to members before the meeting. Neilson and five others convened a special meeting to have the motions validated retrospectively, but their suggestion was ruled out of order.

At the annual general meeting of 31 March 1945, relations between Whitney and Fargie nevertheless remained cordial. Whitney moved Fargie's nomination as president, while Frost and Neilson were re-elected as vice-presidents. Whitney was confirmed as deputy chairman, indicating that the club committee considered the motions passed the previous year were of no effect. However, two further items indicated that the committee had changes in mind: first, rules as approved by the committee for a club to be formed as the Tongariro Anglers Club were presented to members; then the meeting voted for the club of that name to be incorporated.

In early 1945 Fargie decided to give the name 'Tongariro' a secure place in history by changing the name to 'The Tongariro Anglers Club' and registering the club as an incorporated society. In June 1945 the Registrar of Incorporated Societies advised Fargie that the name 'Tongariro' was available for use in this way. Fargie (and presumably other committee members) arranged for a special general meeting to be held on 29 December 1945 to vote on a motion to change the name. All use of the name 'Waikato', officially at least, would be ditched.

It was at this time also that the club made submissions to the Honorary Geographic Board of New Zealand that the name Tongariro be preferred to Waikato for the river. The submission was triggered by that mysterious appearance, recounted earlier, of yet another sign at the Tūrangi road bridge alerting passers-by to the fact that the river was called the Waikato, with some unspecified part of it comprising the Tongariro Reach. The sign was followed by an explanatory statement apparently intended to counter anticipated objections.[22]

Before any further steps could be taken, Fargie and other club members were surprised to learn that Whitney had registered an incorporated society called 'The Upper Waikato and Tongariro Anglers Club' in early October 1945. This development came as a shock; members could not recall a proposal to register under that name being made and approved, and were astounded by the move.

Upon investigation, it quickly became clear what had occurred. Whitney, opposed to the formation of the Tongariro Anglers Club, had let his zeal for the Upper Waikato River name override his judgement. In the course of registering the UWTAC he made what was almost certainly a false statement to the authorities. Legal formalities for the registration of the club included a statutory declaration that Whitney had to make, stating that a majority of members had consented to the club's incorporation. The declaration was completely at odds with the facts — a point Whitney later conceded to the Registrar of Incorporated Societies.

The committee informed the Tongariro Anglers Club of the situation. To provide evidence that there had been no consultation with members, each person received for signing a pre-printed form stating that there had been no notification of registration of the new club and the member concerned would have opposed any such action. Ultimately, the Upper Waikato and Tongariro Anglers Club was de-registered in late February 1946, having had a short, inactive life of some five months. The Fargie faction proceeded to register the name Tongariro Anglers Club later that year. However, that setback did not deter Sir Cecil from trying unsuccessfully to sidestep the name issue with a motion to alter the club's name to Turangi Anglers Club.

The rules for the abortive Upper Waikato and Tongariro Anglers Club reveal that it was very much the creation of Cecil Whitney and those within his circle. These people included Harold Barrowclough, a future chief justice. Indeed, the rules exhibited some unusual features. After stating the club's first objective was the advancement of the sport of freshwater fly fishing, express provision was made for the name of the river. The name Waikato would apply to the course of the river from its source until a point at Waddell's Pool, nearly 18 kilometres from the delta. From that point down to the lake the river would be called Tongariro. This use of two names was clearly based on the earlier decision reached by the Honorary Geographic Board.

TALTAC formed

Several years later, a group decided to form a fishing club for the Taupō region. In late June 1950, 38 people met in Taumarunui with the aim of 'forming an association of Anglers who visit Lake Taupo and adjacent rivers whose voice would carry some weight in obtaining improved conditions in the interests of all Anglers'.[23]

So the Lake Taupo and District Anglers Association was duly formed under the guidance of a committee and a president, Harold Digby-Smith. The association then began to grow. In early September 1950 there were 59 financial members, but at the date of the first annual general meeting in July 1951 numbers had increased to 163.

The committee was quick to meet the Minister of Internal Affairs, the Hon. W. A. Bodkin, to discuss issues concerning the Taupō trout fishery. Items raised included alterations to the dates for the closed fishing season (still in operation in the early 1950s), access to rivers across Māori land, motor-car roads leading to fishing pools, erection of footbridges over the Tongariro River, poaching of trout, the appointment of honorary rangers, research into the fishery and angler etiquette.

Digby-Smith was a good choice for the association's president. As a former public servant (he had been Commissioner of Pensions) he knew his way around the corridors of power, and the deputation was well received. Sir William expressed support for the association's objectives and helped it find land at Tūrangi on which it built its clubhouse, which was officially opened on 6 March 1952.

Footbridges for anglers had been a longstanding issue, because the only bridge on the river was at the village of Tūrangi on the highway leading to Taupō. Anyone wishing to fish the upper river from the true right bank had to make the crossing in a small boat or risk drowning by wading at the tail of a pool. Neither method was an easy task in the Tongariro's turbulent water.

In 1935 the Upper Waikato and Tongariro Anglers Club lobbied for a footbridge to be built near the hatchery camp. The Prisons Department, which owned land on the far (eastern) bank was strongly against the plan, as a bridge would potentially provide an escape route for inmates.

By late 1954, the association's numbers had swelled to 284. The committee decided in June of that year to approach the Tongariro Anglers Club about a possible merger of the two bodies, with the aim of creating one organisation to represent Taupō anglers' interests. The Tongariro Anglers Club accepted the offer, new rules were drafted, and in March 1955 the Tongariro and Lake Taupo Anglers Club (TALTAC) came into being, with 500 members. It is an incorporated society and has been active ever since.

TALTAC's primary aims are to foster and promote angling in the Taupō district, to improve conditions for anglers, and to co-operate with the agencies managing the fishery. When the Tongariro Power Development scheme was built, the club lobbied for measures that would protect the fishing. *TALTAC, The First Fifty Years of the Tongariro and Lake Taupo Anglers Club* recounts the history of the club and its activities.

TALTAC maintains a neat and tidy clubhouse in Tūrangi, right beside the river, complete with accommodation and cooking facilities. Such is the superb reputation of the Tongariro River that TALTAC has some members who live overseas and return each year to fish.

The two trophies which Cecil Whitney (later Sir Cecil) donated when the club was formed are still awarded each year. There are other competitions for the largest fish caught by a junior angler and by a female member, and for the trout with the best condition factor (both rainbow and brown). Not all anglers see fishing as a club or group activity, but contact between TALTAC members provides many angling advantages, the main one being information on fishing conditions.

CHAPTER 13:

The dry fly revolution

'. . . it marks the most momentous angling discovery of recent years.'

— F. E. THORNTON, *THE NEW ZEALAND HERALD*, 31 MARCH 1934

FROM THE VERY EARLIEST DAYS of the Taupō fishery, a wide range of angling methods was permitted on the Tongariro River. Spoon, minnow and artificial fly were all allowed — in a nutshell, the rule was anything except live bait.

Although New Zealand fly patterns were developed over time, the first anglers to tread the banks of the Tongariro used salmon fly patterns transplanted from their UK home. The river flowed deep and broad and its fish were salmon-size, so it must have seemed logical to fish with such standard flies from 'Home' as Silver Doctor, Bulldog, Jock Scott, Thunder and Lightning, and Butcher.

It was also necessary to fish deep in a pool to get results. The correspondent 'March Brown', writing in *The New Zealand Herald* in 1923 advised the use of lead to get the fly well down, adding that he had never heard of a trout being caught with a fly fished at the surface.[1]

Even though the large wet fly fished deep was the prevailing angling orthodoxy, there were occasional exceptions which showed that Tongariro rainbows would come to a surface fly. On a calm, bright day in May 1927 an English angler, N. F. Bostock, took eight rainbows from the Log Pool by fishing a small, lightly dressed fly just under the water surface.[2] Bostock was applying the technique that his friend A. H. E. Wood had used for catching salmon on Scotland's River Dee.

Alan Pye

The Tongariro River was the first Taupō waterway to see dry flies used to tempt the rainbows and browns. In 1924 Alan Pye, an unpredictable Irishman, demonstrated the technique on a small King Country stream to fellow anglers Trevor Withers and Frank Yerex. Pye's Waikato River fishing lodge business was some years in the future, but even then he was a superb fly fisherman. After a weekend of tuition from the master, Withers and Yerex went back to the Tongariro and took trout there with surface flies.[3] But despite their success, dry fly fishing failed to take hold, and the downstream lure remained the only method on the river; probably because in the mid-1920s the second golden era of the Taupō trout fishery had arrived and the fish were at their heaviest since the great days before World War I.

So it was not until the early 1930s, when fly fishers were confronted by low river levels, that the dry fly came into its own. In those conditions the Tongariro trout became shy, and anglers' large wet flies proved to be almost useless in the very clear water. It was around this time that Alan Pye discovered the massive caddis fly hatches on the Waikato River and moved to set up his fishing lodge on its banks. Pye studied the natural insects closely and devised patterns which proved very attractive to the trout. Before long this Wizard of the Waikato, and his by then established Huka Lodge, were attracting anglers from all over the world.

Joe Frost

The man who can be credited with firmly establishing the dry fly method on the Tongariro River is Joseph Colston Frost, an Englishman born in Herefordshire's Wye Valley in 1893. Frost's father was a keen angler, and as a lad Joe Frost learnt to fish in Ireland for trout, salmon and coarse fish. By the time he reached his teens he was a skilled angler and fly tier.

He was already in the British army when World War I broke out, having enlisted in 1910 with the Royal Warwickshire regiment, but was discharged in 1916 on sickness grounds. While recovering in an army hospital, a New Zealand sergeant told him about Taupō's fabulous fishing waters. Frost was fascinated by the tales of the Antipodean monsters, but several years passed before he made the trip, in late 1923.

On arrival in New Zealand, he undertook a sort of angling survey lasting some two months. During that journey around the North Island he first encountered the Tongariro River, opening his account with a superb 14lb rainbow from the upper reaches.[4] That catch (which was to prove his largest) was the start of a very long association with the world's finest rainbow trout fishery. He is considered to be the first angler to fish nymphs and dry flies on the Tongariro.[5] While Frost was an early convert to the Tongariro River fishing, he was too late to enjoy the first boom period enjoyed by Major Rhys Wykeham Jones, William Soltan Pillans and A. G. Campbell when trout as large as 20lb were caught.

Joe Frost's first New Zealand business venture was as a butcher in Wellington. The shop failed in the late 1920s, apparently because its owner was spending too much time on fishing and not enough on the needs of customers, although the onset of the Depression can hardly have helped. He must have then decided to make his hobby his work and relocated to Taupō. After setting himself up at Tokaanu with a house and a job at the local trout hatchery, Frost turned his energies to the Tongariro trout and their diet. A skilled fly tier, by the early 1930s he was making and selling nymph patterns which closely imitated the aquatic insects on which the trout fed. There were two main flies he found successful; one with an olive green body, the other brown. These he fished near the surface on a leader greased to within 4 inches of the fly.

Frost spent three seasons working at the government trout hatchery. Hatchery work occurred in the winter (closed season) period, so when that work slowed up Frost was able to earn income as a gillie attached to Taylor's Lodge, hard by the Island Pool.

Frost's guiding business

Frost began guiding trout fishers on the Tongariro by accident. Two English anglers staying at a local fishing lodge had fished for days with no luck when they met Frost on the upper river one evening. Observing his fine casting and impressive catches, they beseeched him for advice on how to catch a trout. Frost agreed to show them the Tongariro technique, and under his tutelage over the next few days both men caught fish, to their great delight. Frost was equally delighted with the £10 they paid him for his services.[6]

As a guide Frost could earn in one hour what he had previously earned in one day as a gillie,[7] so after a short time his capital built up.

Thanks to a large order for flies from the tackle dealers Tisdall, Frost gave up hatchery work and trebled his earnings. Soon he bought a Tūrangi cottage, and in 1929[8] he set up his own business as a tackle retailer, fishing guide and fly tier.

Overseas anglers, especially those from the chalk streams of England, found the rough-and-tumble dry fly fishing on the Tongariro a new experience. There was no waiting by a placid pool for a trout to rise; instead, Frost placed his customers in rapids and told them to fish the pocket water using a short drift. The snout of a weighty brown or rainbow would then appear, the angler (if alert) would tighten, and the fight would be on.

Frost's other strategy was to fish slower stretches where the overgrown banks made it difficult to use the traditional downstream wet fly method. [9] Dan's Creek, a side channel of the Tongariro near the delta, was a favoured spot. Stalking quietly upstream, Frost would spot fish lying near the shelter of overhanging branches. Again, a short but accurate cast was all that was required.

Dry fly fishing takes hold

News of Frost's catches began to appear in the metropolitan daily newspapers, which featured regular Taupō fishing reports. In February 1934, *The New Zealand Herald* noted that dry fishing activity was becoming more common, with Frost taking two trout weighing 7½lb each from the Judge's Pool one evening.[10] The year before, Frost had written to F. E. Thornton, the editor of the *New Zealand Fishing & Shooting Gazette*, advising that in May 1933 he had taken some good bags on the upper Tongariro pools using a dry fly on a light rod. Thornton was an experienced exponent of the dry fly art, having fished at Alan Pye's Huka Lodge, and had already tried fishing a dry fly on the Tongariro, but the small, feathered offerings seemed out of place on the river's roiling rapids and swirling eddies; he simply could not countenance that type of fly fishing on such boisterous water.

Frost's reports convinced Thornton he should try again. In late summer 1934 he booked in at Taylor's Lodge and took to the water:

> . . . I strolled down to the famous Judge's Pool behind the Lodge with a 9 foot Makuri rod, a line that matched it to perfection, a 2x cast, and one of the singularly beautiful dry flies built by Allcock's . . .[11]

Immediately, large trout rose to his fly, although he missed the first seven before a fat 5½lb hen fish stayed connected. Others followed with the evening rise, so by the end of the day Thornton was completely convinced the dry fly was lethal in the right conditions. He then began to experiment with different-strength gut casts, fishing a Red Sedge fly designed to imitate a natural insect he had observed coming off the water.

The new religion began to spread. The day after Thornton's success he schooled a visitor from England, Hulda Whyte, in the new art. Hulda Whyte had caught salmon on the Thurso using large rods, but had never fished the dry fly. After Thornton and a friend gave her a demonstration, she caught trout from the start and left her salmon rods untouched for the rest of her stay.

Fighting big rainbow trout in heavy water was new and exciting for Thornton:

> This was an amazing experience for me. From these deep swirling pockets, a huge rainbow would rush the dry fly (I was down to using a No. 12 Red Peveril) from a depth of five or six feet, and give a wonderful exhibition of leaping. With these small hooks and the big belly in the line, when they turned upstream in heavy water if a fish leaped again, as they often did, the hook would tear out . . .[12]

The whole trip became the stuff of dreams. Every day trout fell to his casting. Once Thornton learnt to read the water and found the best lies in each pool, it was merely a matter of placing the fly accurately and taking the fish.

Thornton's tackle was a heavy dry fly rig; necessarily so because he was fighting big fish in a fast-flowing river. It performed admirably. With his 9-foot rod (made by the New Zealand firm W. H. Tisdall Ltd, weighing a little over 5oz) and Hardy's Uniqua reel, he took more than 100 trout in three weeks. Sixty of those fish weighed 5lb or more. At least 100 yards of line was essential, as the first wild run of a hooked fish would strip 50 yards of line.[13]

The flies he used included the Brown Drake, Hardy's salmon dry flies, and Brown Sedges tied by Joe Frost, who by then was working as a gillie at Taylor's Lodge.

Captain Arthur Richardson, who had helped develop dry fly fishing on the Waikato River, was equally enthusiastic: 'The Tongariro is not a chalk stream, its [sic] only some of the finest dry fly fishing in the world.'[14] Richardson landed trout up to 8½lb from the Tongariro's middle pools, but frequently the fish won the battle. On two consecutive afternoons he hooked some 30 fish, but failed to land any. The No. 10 hook simply would not hold in the trout's jaw unless the strike was made from a position directly behind the fish. When he spent 10 days fishing at Tūrangi in March, Richardson brought only his dry fly rods with him — testimony to the effectiveness of the technique.

Joe Frost's dry flies

Frost persisted with his efforts, learning as he went. A year later, his catches for one week in March 1935 included a fine brown trout of 10lb from Dan's Creek on the dry fly, and 11 fish caught in the upper river pools over two mornings using a large Brown Sedge (caddis) fly.[15] By this time Frost was advertising his services as a teacher of the dry fly technique. The reference on his sign to 'tuition in dry fly fishing' is significant: he would not have bothered to give instruction in an angling method that did not give results and for which there was no demand.

The dry fly technique was partly dictated by the time of year. Frost noticed that in November the Tongariro rainbows gorged on green beetles which inhabited the mānuka bushes fringing the riverside. Using a beetle pattern, he took many fine specimens. But he also found that the fish would come readily to a very large dry fly — what he described as a feather duster — tied on a No. 10 hook and fished blind. In the evenings, the rainbow trout seemed to be more discriminating and a near-exact copy of the natural beetle was required.[16]

Another dry fly pattern Frost devised was the Prismatic; so-called because it had two hackles, one brown and one Cambridge blue. When sunlight shone through them, these colours in combination gave a prismatic effect similar to the translucence displayed by an aquatic insect's wings.[17]

Because most Taupō waters are now open for fishing all year round, it is easy to overlook the limited fishing season which prevailed in those times. The season for the Taupō district opened on 1 November and finished the following 31 May. That meant the bulk of the season covered spring, summer and autumn, with the fish entering the river from about March onwards. In retrospect, a large part of the season was in the warmer part of the year when dry fly fishing would have been viable.

Joe Frost with two satisfied clients and their catch from the Tongariro River, 1936.
WA-03145-F. Alexander Turnbull Library, Wellington, New Zealand. /records/30653029

A national pictorial, *The Weekly News*, often carried Taupō fishing reports, and Frost used these to inform readers of the Tongariro dry fly fishing. He never wrote articles under his own name (it seems the pseudonym he used was 'Red Sedge'), but he is listed in the regular reports of anglers' catches.

With his deep knowledge of Taupō trout, Joe Frost was able to defend the fishery when it was occasionally criticised by foreign anglers who felt that they had not got the results the tourist advertisements promised. One disappointed visitor was Sir Walter Windham, a pioneer aviator who came to Tūrangi in early 1934. Although the Tongariro River was low and clear, Windham refused to take the advice of local anglers and fish the dry fly (the New Zealand patterns he thought were unbalanced) and persisted with large lures fished downstream. Joe Frost went with Sir Walter several times and noted that it was the New Zealand flies that caught fish.[18]

Frost's tackle business suffered during the years of World War II and even after. Import restrictions imposed by the government meant that fly rods and reels could not be brought into New Zealand. Frost got around this difficulty by working his passage to England on a ship, where he arranged to act as New Zealand agent for leading tackle makers with whom he had been dealing in pre-war years.

Joe Frost added to his growing reputation among New Zealand and overseas anglers by lending his name to several split-cane fly rods the English tackle makers J. J. S. Walker Bampton made to suit conditions on the Tongariro River. For dry fly work there was the J. F. Lightweight, 8 feet 6 inches long and weighing 6½oz, and the J. F. Tonga, a little longer at 9 feet 3 inches and 7½oz. For long casts on large rivers the Joe Frost 'Tongasand Special' was an 11-foot, three-piece rod weighing 12oz; presumably for use with a wet fly.

With the guiding and fishing tackle business back on track, Frost decided to move on. In 1949 he sold to Geoff Sanderson, but remained living in Tūrangi until 1958, when he moved to the South Island, settling in Timaru and starting a new phase of his fishing career in brown trout country. (His private life is somewhat obscure, but because he married his third wife in 1958 it is not a stretch to suspect that — following a route sadly not uncommon with keen fishers — fishing came first and his domestic relationships suffered as a result.) After his shift south, the Tongariro River clearly still kept a hold on him, though, as he could often be found writing to government ministers with ideas on how to manage the fishery.

When Joe Frost died in 1983, no monument was erected at Tūrangi to mark his passing, nor does a pool on the Tongariro River bear his name. Fishing tackle collectors may obtain a split-cane fly rod with his name on it and wonder at it. Nonetheless, the significance of his contribution is enormous, for since the 1930s dry fly fishing has continued on the river, principally being used in summer. When fishing to sighted trout it is hugely exciting, but today it tends to take a back seat to wet fly and nymph. However, the large brown trout that inhabit the lower reaches of the Tongariro River offer a marvellous opportunity for the dry fly enthusiast, and indeed there are some anglers who specialise in targeting these challenging fish.

CHAPTER 14:

Tongariro camps and lodges in the interwar years

The Tongariro is undoubtedly the anglers' Mecca, and its fame is now worldwide.

— ***THE NEW ZEALAND HERALD*, 20 OCTOBER 1931**

THE 1926 LEGISLATION OPENING UP ACCESS to the waterways was a major threat to the fishing business Robert Jones had operated for so long. The change in the law effectively made Taupō riverbanks accessible to all anglers holding a valid trout fishing licence for Taupō waters.

The 1930s were also marked for camp and lodge owners, as for all New Zealanders, by the effects of the Great Depression. The economic downturn — largely caused by a drop in New Zealand's export receipts — brought high unemployment, soul-destroying make-work relief schemes, and cuts in wages of civil servants. National morale fell and was accompanied by a mistrust of the government; attitudes that were not lifted until the election of a Labour government in 1935.[1] By that time economic recovery was well underway. But the call of the river is strong, and many people went to the Tongariro River to fish and forget their worries, allowing those offering accommodation to anglers to survive.

Delta Camp

The government plan was for Tongariro pools formerly reserved for guests at Hut Camp and Delta Camp to be open to any angler provided he or she stayed within the 1-chain (22-yard) strip reserved by the law along the water's edge. Temporary camps could also be set up within that strip. One government official commented: 'We will thus have complete control of the river and its banks, notwithstanding the fact of Mr Jones' camp.'[2]

It became clear that the Jones camp at the delta had been singled out for special treatment. The island on which Delta Camp was located ('Jones Island') and other islands at the Tongariro mouth were not deemed to be Crown land. Secondly, anyone holding a Taupō district fishing licence was given a general right of access over the south-western corner of the island and along a narrow strip (extending 7 yards from the water's edge) around the rest of the island.[3]

The following year, 1927, the government issued regulations authorising permanent camps to be sited on the Taupō rivers, provided an annual fee was paid.

For people who could not obtain a bed at the hotel or a lodge, the only alternative was to camp out under canvas. Jellicoe Point and Mission Bay on the eastern side of Lake Taupō were the most favoured locations. These camps had no running water or sanitation, and conditions became unpleasant when many anglers were in residence at peak holiday times.

The 1927 regulations also allowed riverside camps, subject to rights for other anglers to pass on the adjacent riverbank to fish. This meant a group of anglers could not take de facto control of a pool for days at a time. Although the income from his fishing camps was seasonal, Jones must have earned enough additional income from the hotel business to keep his wife and three children. When he came to take over the Tokaanu Hotel at the start of the twentieth century, he brought years of farming experience and a reputation as one of the best horsemen in New Zealand. In the days before the motor car, such skills would have been essential to anyone running a fishing camp in a remote area like the Taupō region.

The Delta Camp business carried on during World War I, although the conflict made it more difficult for Northern Hemisphere anglers to visit. By 1921 the charge for staying at the camp was £1 per day;[4] presumably on a bed-and-board basis (food provided) — a small charge for what was some of the best fishing in the world.

As the Tongariro's fame grew, the camp facilities were improved. By 1926 Delta Camp had eight tents with wooden floors for anglers' use. A wooden homestead housed the two camp staff, and also contained a dining room for eight people, with a smoking room attached. A magnificent garden containing plum, pear and walnut trees planted by the early missionary John Grace provided a colourful backdrop.[5]

While the trout fishing on offer may have been the best in the world, the same could not be said of the camp accommodation. Visiting anglers from the United Kingdom and the United States were taken aback at the basic nature of Delta Camp facilities in 1927: a tin bucket to wash in and one towel a week for 30 shillings a day board. As one American pointed out, the visitors might be rich, but they were not fools and could be stung only once.[6]

Robert Jones died on 31 May 1924. When accounts of the estate were tallied, as we learnt earlier, it was revealed that many local Māori were in debt to him — a consequence no doubt of Jones's willingness to provide credit.[7] Government social assistance programmes under the first Labour government were still some years away, so Jones had been providing a form of public welfare to the community.

Jones's estate was a valuable one. As 'King of Tokaanu', his business included three fishing lodges, the Tokaanu Hotel (the only licensed liquor outlet in the district), a general store and a billiard saloon, as well as the provision of services as a carrier and baker.[8] Nothing happened in Tokaanu without Jones getting a slice of the pie.

Jones's brother-in-law, John Asher, ran the hotel, store and fishing lodge businesses on behalf of the estate trustees from 1924 until 1932. In that year he induced the trustees to lease almost all of the estate's assets (largely the hotel and general store) to him for five years with an option to take up another five years. Instead of managing someone else's property, Asher was now legally managing his own.

By the start of the 1930–31 season, Asher had established a new camp by the Tongariro River road bridge. Advertisements referred to this camp as Asher's Tongariro Fishing Lodge ('under Royal and Vice Regal patronage'). He claimed to provide his guests with private fishing water in the Tongariro River. The Tokaanu Hotel and Delta Camp were also advertised in newspapers under the name of 'J. Asher, proprietor'.[9]

These dealings made Asher the new 'King of Tokaanu', but the operation was not especially successful, probably thanks to an economy burdened by the 1930s Great Depression. Although the

Tokaanu Hotel and store had year-round custom from locals and tourists, the fishing lodges could operate only during the fishing season, which began in November and ended the following May. Debts continued to burden the estate, and by 1935 creditors were threatening a forced sale of the property.

Debts were not Asher's only concern. History tells us that doing business with close family members is never sound practice, particularly when one person has complete control of the assets, and so it proved for Robert Jones's widow, Annie Jones. John Asher, by mingling his duties as a trustee with his own business affairs, put himself in a position of divided loyalties and conflicting interests, with the inevitable result. Asher was replaced as trustee in 1929, but resumed control of the estate assets in 1932 when he became lessee of the hotel and store businesses and acquired the leases of the fishing camps. This effective usurpation was a step too far for the widow.

Annie Jones sued Asher in the 1930s, seeking possession of the Tokaanu Hotel, the store and the Tongariro fishing lodges. She alleged he had mismanaged estate assets, and not paid the rent due under the leases, which he had induced her to grant him. The case was messy and inconclusive, because the business books had become muddled, with some of Asher's expenses commingled with those of the estate. At the same time Asher had given useful service by arranging with the tribal trust board to pay a sum to clear the debts which local residents owed to the general store.[10]

In 1936 the court ordered that Annie Jones was to have possession of the hotel and store, but left the leases for the fishing lodges in Asher's hands. The judge directed that accounts be taken to clarify the conflicting financial claims, which he described as 'this mass of confusion'.[11]

The affair came to an untidy end when the court cancelled the hotel lease to Asher for non-payment of rent. That order was later reversed when the figures were re-checked.[12] The arrangements Asher had made concerning the fishing lodges were left untouched. Asher had given evidence that he held the Delta Camp lease directly from the Māori owners and was also lessee of a camp in Tūrangi.[13] Whoever was to blame, an estate worth £20,000 on Robert Jones's death had been allegedly reduced to £9,000 a bare 12 years later.

Amid the turmoil, Annie Jones tried to lease the Tokaanu Hotel to Thomas Hurley in February 1936, but the deal was stalled by the protracted legal proceedings over the property in Robert Jones's

Delta Camp, 1930s.
From the collection of Mrs E.G.Renz

estate. Hurley finally got the lease transferred to him in 1938 after the police complained to the liquor licensing authorities that Asher was not running the hotel in a proper manner. Evidence of mismanagement included allegations that a visiting football team was supplied with alcohol on a Sunday.[14] Asher denied the charges, but was ordered to transfer the hotel licence to some more suitable person. Hurley held the business for several years before selling it to the Tourist Hotel Corporation in 1944.[15]

In the 1930s, Delta Camp was managed by John Selby.[16] The last owner of Delta Camp was James Scott Wilson, who had charge of it from the start of the 1938–39 season until 1941. The fishing was affected when the level of Lake Taupō rose permanently after control gates were installed on the Waikato River, the only outlet from the lake, for power generation purposes.[17] The lake rose by 3 feet in October 1941, flooding the camp cottages.[18] The government of the day compensated Wilson for the financial losses. It appears that John Asher was still active in the area at the time, though, as he was one of several persons who similarly sought government compensation for economic loss caused by the lake's raising.

Hatch's Camp

While Robert Jones might have been the key player in Tokaanu, he didn't have the area completely to himself. In mid-1928 George Hatch began preparing a fishing lodge in time for the 1928–29 season. This camp occupied a prime spot on 2 acres of land above the Tongariro River bridge — the site later occupied by Taylor's Lodge beside the Island Pool.[19] Campsites were also available for anglers at the Department of Internal Affairs hatchery, several hundred yards downstream from Zane Grey's campsite at Kowhai Flat.

Hatch's Camp was in the same parcel of land containing a property owned by Judge Ostler, that well-known angler and hunter we encountered in Chapter 9. There were huts, each with a bathroom, and anglers took their meals in the camp dining room. As an added service Hatch ran his guests to and from the Tongariro River pools at no extra charge.[20] That was a smart move on his part, motivated perhaps by complaints from overseas anglers that other hotels overcharged them for the privilege of being conveyed up and down the river. In these early days, of course, motor cars were still a rarity in Tokaanu and visitors were at a disadvantage. To give his lodge added appeal, Hatch also supplied liquor to his guests in flagrant breach of the law; an offence for which he was convicted and fined in May 1929.

In the Tongariro's lower reaches, the riverbank was covered in mānuka trees and teemed with furred and feathered game. The thick vegetation provided ideal cover, so Hatch advertised it as a shooter's paradise for those seeking duck, swan, quail, hare and pig.

As noted in Chapter 9, once the agreement between the Crown and Ngāti Tūwharetoa had been made, the Department of Internal Affairs staff began to provide better access to the Tongariro River for anglers. A motor road running parallel to the river was prepared in late 1926, opening up 15 miles of water, although five miles of riverbank remained in private hands under European title.[21]

The department also cleared riverside foliage to make room for anglers' campsites. These were available for groups of up to six people at a charge of 10 shillings per month.[22] While the campsites meant anglers were never far from the river, as we have seen the occupation right was not an exclusive one, with the campers being required to allow other anglers to pass through their camps.

The 1926 access agreement was a major shift in policy, effectively creating the public fishing access that all Taupō anglers enjoy today.

Tūrangi township, with the Tongariro River. (Aerial photograph of the Island Pool, with Taylor's Lodge and store in foreground).
Ref: WA-25094-G. Alexander Turnbull Library, Wellington, New Zealand. /records/22861451

Nevertheless, Asher's camps continued to provide accommodation as before, principally to the overseas anglers who stayed in Tokaanu for lengthy periods. Previously vested interests were not going to relinquish their livelihoods without demur. And, to be sure, these interests were at least in part officially acknowledged, with the owners of these camps notably described in print as the previous controllers of the Tongariro River.[23] Still, to some degree their activities now came under the purview of the department, an incursion that was far from welcome, and in some cases contested. For example, in 1928 Asher tried to dispute the fees he was asked to pay for having permanent huts at Hut Camp within the marginal strip on the Tongariro River. The government refused to accept his argument that the riverside reserve strip should be taken as starting at the water's edge, an interpretation that would mean his camp buildings were located outside it.[24]

Taylor's Lodge

In late 1930 an advertisement for 'The Tongariro Fishing Camp and General Store (late Hatch's Camp)', now under entirely new management, showed that the lodge had changed hands.[25] George Hatch had not made a success of the business, and he was declared bankrupt in mid-1930. The new owner was James W. K. Taylor, who upgraded the facilities in several ways, including the installation of hot and cold showers. Taylor's Lodge would become a Mecca for Tongariro fly fishers in the years before World War II. In 1936 James Taylor sold the business to R. E. (Bob) Cuthbertson. The lodge also fulfilled an important function by housing the Tūrangi Post Office. In 1939 a young woman, Alpha Jane Smith, arrived at Tūrangi to take up the position of postmistress. She found the village had no police station, electricity, church, doctor, bus service, school or piped water supply. At the end of the war, despite a considerable difference in their ages, she and Bob Cuthbertson married and kept the fishing lodge business going for another 20 years.

Taylor's Lodge employed Joe Frost, a fly tier, as a guide. It also had several governors-general as guests and hosted meetings of the UWTAC.

Hatchery huts

Once the Tongariro hatchery was established in 1927, the Department of Internal Affairs established four huts there for anglers to use. These were hard by the Birch Pool, and comfortable if a little spartan. Situated among mānuka trees, each hut had bunks, a table, stools, rudimentary shelving, a washstand and an outside fireplace.[26] These huts were used by anglers for decades and offered a more affordable alternative to the lodges.

Fishing trip costs

For New Zealand anglers the most popular times to fish at Tokaanu were at the start of the season (1 November), Christmas/New Year and Easter. Overseas visitors, having travelled a long way by ship to reach New Zealand, tended to stay for longer. Many were members of the titled classes who had no need to work for a living, and they would include the Tongariro on their list of fishing waters before making their way down to the brown trout waters of the South Island.

With a little careful planning, overseas visitors could have a relatively low-cost fishing trip. Two anglers, C. Stockdale of British Columbia and B. H. Smith of England, arrived in New Zealand in January 1928. They bought all their fishing tackle in Auckland, together with a bulk order of food and other provisions at wholesale prices. At the end of the trip they calculated that — after factoring in costs of travel by steamer to and from New Zealand, food, bedding and other items — their daily expenditure was £1 per day.[27] No other country, said Stockdale, could offer such a variety of fishing and hunting to overseas visitors at such a low cost.

Despite the difficulties in getting to the Tongariro, it began to be very crowded at holiday times. The most popular pools, the Hut, Jones and Log pools, were those downstream from the main Taupō–Tokaanu road. The upper pools, such as Major Jones and Duchess required motor cars for access unless the angler was prepared for a long walk. It must be remembered that in the early days of the fishery, it took a long trip over unsealed roads to reach the Taupō district. Bad weather would often make roads impassable, so the ordinary angler unable to spend weeks or months on the river could realistically fish only at Easter or Christmas. At those periods the numbers of rods on the river peaked and people found it difficult to wet their lines.

The fishing camps and lodges came to meet a permanent demand for accommodation by anglers in the years before World War II. It was the practice of people like J. S. W. Neilson and B. U. Barlow to return to the Tongariro River year after year like migrating birds. Taylor's Lodge also became the venue at which members of the Upper Waikato and Tongariro Anglers Club held their annual general meeting and presented fishing trophies.

Tourist numbers grew in the years after the Great Depression as New Zealand came to be seen as a desirable place to visit. Even so, patronage from anglers was linked to the fishing season, which ran for only part of the year, and so accommodation providers looked to other travellers to provide business during the closed-season months from June to the end of October.

CHAPTER 15:

Three great Tongariro pools

. . . there maybe those . . . who of a winter's evening beside the fire, will live over again their exploits and gossip nostalgically of these waters that have gone forever.

— HAROLD HICKLING, *FRESHWATER ADMIRAL — FISHING THE TONGARIRO RIVER AND LAKE TAUPO*

ONCE THE TAUPŌ RAINBOW TROUT FISHERY was established in the early 1900s, anglers explored the lower Tongariro River and found the best fishing locations. Two pools below the bridge, the Hut Pool and the Log Pool, quickly emerged as truly outstanding, giving up massive trout season after season. Situated close to the Hut Camp, they could be reached with comparative ease, and many anglers considered them to be the most productive pools on the river.

The Dreadnought Pool, higher up the river, also had its devotees, but was harder to access. Of these three only the Log Pool survived the flood of 1958, but even that became less productive.[1]

The Hut Pool

The Hut Pool was a short distance downstream from the main highway leading to Taupō. To reach it anglers took the road to Tokaanu and turned down a side road which led to Hut Camp. There, they would find a gently curving stretch of water some 600 yards (550 metres) long from the rapids at the head to the tail.

According to Ronald Cumberworth, the centre of the Hut Pool had moderately fast water which the angler could fish by wading along a shallow ridge of boulders in mid-stream. Trout would lie on each side of the ridge. Further down, where the pool widened and slowed, there was room for four or five double-handed rods to explore the deep water along the eastern bank.[2]

Anyone wanting to fish the shallower stretch known as the Nursery had to wade 100 yards across the tail of the Hut Pool, but the effort was well worthwhile.

Bob McDowall, one of New Zealand's finest fisheries scientists, learnt to fish the Hut Pool in the 1950s as a lad. Years later, in the twilight of his life, he recalled the experience in *Memories of an Antediluvian Tongariro Fisherman*:

> . . . the bigger water of The Hut Pool was fished from the left. It was narrow and deep at the top (like any pool), making for an easy cast across the river — not unlike the top of The Major Jones today, though somewhat wider. And as you worked your way downstream it became both wider and deeper (again like any pool!), and in the heart of the pool it was a long cast across to the manuka-clad right bank. This tempted anglers to wade out as far as possible, and only the big men could cover the whole pool. Wading well out was hazardous, as there were big boulders here and there, and these had to be circumnavigated with considerable watchfulness if the swift flows were not to knock you over, pick you up and carry you off. So, you'd edge downriver foot by foot, feeling for the big 'ghoulies' [boulders], and circumnavigate them with real care. Well downstream a torrential rapid swept away to the left into The Boat, which was a short, rushing pool that took about a third of the river's water. It occasionally held fish, and sometimes we would be tempted to move down there — usually only when nothing was doing in the main pool. To get to The Boat you had to cross the head of that branch in the lower Hut Pool and then work down to fish it from the right bank. It was difficult wading, and I can still sense the terror of being washed off balance, and going down to wrap my sodden arms around a large boulder while I repositioned my feet (and my dignity), and slowly edged my way back to the bank, fortunately still with my rod clutched in my hand — though I don't know what instinct made me hold on to it.

The Boat didn't seem to hold fish much. The bottom of The Hut Pool was known as The Nursery, and here there was a wide expanse of gently flowing water which eventually split into several small channels. It was a metre or two deep and was my favourite part of the pool. In those days, before there were sinking lines, it was really difficult to get your line down around the bottom. This was then a general problem never experienced by those who haven't fished with silk lines — and for those who haven't,

Fly fishers at the Hut Pool, circa 1941.
Auckland Libraries Heritage Collections 1370-626-11 Photographer W. B. Beattie, *The New Zealand Herald*.

Trout fishing, Hut Pool, Tongariro River.
Ref: WA-09627-G. Alexander Turnbull Library, Wellington, New Zealand. /records/22628168

> there is still no line that handles as beautifully as a silk one, but they didn't sink well and did rot, eventually, no matter how much you cared for them.[3]

John Clemance, a former president of the Tongariro and Lake Taupo Anglers Club, first fished the Hut Pool in 1950. On such wide water there was room for plenty of rods, and on occasion he counted 17 anglers fishing from the left bank while another seven cast from the right.[4] He recalls a signed and dated statement on the wall of an old accommodation unit in the Bridge Lodge which certified that at least 256 trout were taken from the Hut Pool on 14 May 1949.[5] That may explain why he considered it to be incomparable fishing water.

Many people who fished the Hut Pool became so devoted to it that it was the only pool they fished. And once in the river they stayed there; Harold Hickling, as a newcomer to the river in the 1950s, likened the long line of fly fishers standing in the water to ships' mooring buoys.

In 1940, during the last fortnight of the season, one angler took 80 trout from the Hut Pool in 11 days, scoring a limit bag on three successive days.[6]

Log Pool

For trout migrating up-river from the great Taupō lake, the quiet waters of the Tongariro's lower reaches at first were easy-going. The waters of the delta were broad and shallow, with quiet willow-fringed stretches that gave no hint of the turbulent pools that lay ahead; the river's silvery thread spreading out on the flat land and dividing into several strands as it neared the end of its quest.

After several kilometres, the river narrowed and deepened to form what would become one of the most famous and revered pools in the world of trout fishing. This stretch was the Log Pool, formed after the river divided at Cupples Island to make a long and turbulent rapid on the eastern side. This stretch created perfect conditions for the angler to seek the trout holding in the faster water. Anglers would come from all over the world to fish it with fly and minnow, many standing in line to wait their turn to hook and then battle its large trout.

By the early 1920s, reports on the Tokaanu fishing invariably listed the number and impressive weight of Log Pool catches. Perhaps because its fishing was now being ranked with that provided by the

The Log Pool, March 1944: deep, wide water near the lower half required long casts to cover the fish lying near the far bank.
Joyce Galbraith.Supplied by Taupō Museum and Art Gallery.

great Norwegian salmon rivers,[7] newspapers would refer mistakenly to the Tongariro's 'salmon trout'. Tongariro regular and author Harold Hickling records a catch of 11 fish by George Cotterell in May 1924, ranging in weight from 10lb to 16½lb.[8] That same month, F. A. Fullerton-Smith caught two huge rainbows from the pool, weighing 19¼lb and 17¾lb.[9] In 1927 it was the natural choice for visiting royalty, the Duke and Duchess of York, to cast a fly.

The potent force of attraction exerted by the Tongariro River, and the Log Pool in particular, was such that even during World War I foreign anglers would make their annual pilgrimage despite the disorder in international affairs. Arthur Millington Naylor, having wearied of big game hunting in Africa, ventured to New Zealand to fish in February 1915, together with Arthur D. Shilson and Joseph C. Buckingham. By that time the trout weight had declined, but good catches could still be made.[10]

The uninterrupted flow of tourist anglers is partly explained by the rules of war: although Germany had declared the North Sea a zone in which ships of its military foes would be sunk without warning, the outcry following the sinking of the neutral ship *Lusitania* made Germany change its policy. The submarine blockade was suspended until 1 February 1917, at which time the Kriegsmarine adopted unrestricted submarine warfare.

And the fishing was worth the trip. A visiting Australian angler caught 29 trout in one morning from the Log Pool[11] — a total considered a record — while English angler C. G. Haywood took seven rainbow trout averaging 12lb each in one day's fishing in 1927.[12] To do battle with these monsters in the wide water he used a 10-foot, 8oz fly rod; a light choice in the circumstances, and one which must have been tested to the limit, with a proportion of fish hooked getting away.

The regulations in force at the time allowed anglers to use artificial fly, spoon and minnow, with no restrictions on weighted tackle. To get their lures down near the riverbed where the migratory trout lurked in the deep, forbidding pools, anglers added lead to their lines. This practice made casting more difficult, but the very long rods in use (sometimes 16 feet or more) could accommodate the extra weight.

One method from the United Kingdom was introduced in 1927 by an English angler, N. F. Bostock, who had fished with A. H. E. Wood, the inventor of greased-line fishing for salmon. This technique employed a single-handed rod to cast a small, sparsely dressed fly across and upstream. The fly was allowed a drag-free drift. Bright, sunny days were reputed to be the best conditions for this method. Bostock caught eight trout in the Log Pool in this way,[13] which may explain why he thought the Log Pool was worth all of the other Tongariro pools put together.

In one week in 1924 the average weight of trout caught in the pool was 11lb.[14] Drought conditions several years later saw that average drop by half. The pool's reputation was restored partly in 1928 when in a single day six anglers took 22 fish from it, averaging 7lb.[15]

In the early days the Log Pool could be fished easily from the true left bank. But a riverbed like the Tongariro is never stable, and, after a large flood in 1926 formed an island which divided the flow in two, most of the water feeding the Log Pool went into the larger channel on the eastern side of the river. To reach the fishable water, anglers had to wade a deep, fast-flowing rapid called the Log Crossing.[16]

An old Tongariro hand, revisiting the river after some years' absence, found that big changes had occurred by 1929, but the fish remained:

> . . . wishing to renew our memory of old-time scenes,[we] made our way to the Hut Pool Camp. We were astounded at the changes in the river since our last visit. The Hut Pool itself was the same as of yore, . . . The old-time Parade, that most famous spinning pool, is now called

> the Boat Pool, and in the middle of it there is now a boulder bank island, where formerly we cast our minnows. . . . I decided to try the Log Pool, so made my way to the end of the Parade, and was dumb-founded. The Log Pool had gone, absolutely gone — not a sign of it. Dry land had taken its place, and looking beyond for the long rapid, it had also gone; but in its place there was a broad sheet of easy fishing water, which was subsequently discovered, and was now called the Log Pool, and from which as large numbers of fish are taken as from the old Log Pool.[17]

The *New Zealand Herald*, in its report on the 1933–34 season,[18] stated that the Log Pool 'world-famed for its fine fish' lived up to its reputation and fished consistently. Notable catches were an 11lb male fish caught by Mr H. Willis of Auckland on 1 May 1934. Later that day, Commander E. Rhodes caught a 10lb trout from the same pool — the first time the Tongariro had yielded two double-figure trout in one day. Both fish were kept for mounting and future publicity use.

The previous March, Wellington angler and Tongariro regular Robert Begg had caught the best river fish for the season from the Log Pool — an 11½lb fish.[19]

These catches were made despite the Tongariro being abnormally low for most of the season.

A feature of the river in the days before swing bridges and easy riverbank access was the signposting of places where anglers could safely cross the river. In late 1939 a road was built allowing motor vehicle access to the Log Pool and others nearby. The new road was an improvement over the unsealed track from the highway bridge, which was often impassable by motor cars after rain.[20]

The erosion problem

The erosion problem, which first surfaced in 1926, continued into 1928. By that time, it was feared that the left-side channel would become the main flow, making the fishing water unapproachable from the true left bank. A news report noted:

> It is a pity that this famous reach is disappearing, as the Log Pool has in the past become well known as one of the finest pieces of fishing in the world, ranking for size and weight of fish with some of the celebrated pools in the Norwegian salmon rivers.[21]

The Wanganui Acclimatisation Society had written to the Department of Internal Affairs in 1927 recommending that a groyne should be built above the Log Pool to divert the flow. This letter received the standard brush-off from officialdom: 'the matter will receive due consideration', and by 1930 nothing had been done.[22]

To some degree that inaction must have been caused by the powerful economic and political forces that had combined to drag the international markets into what became the Great Depression. The New Zealand economy, based very largely on agricultural production, went into a downward spiral as world prices for meat, wool and dairy products slumped. When the government of the day, hamstrung by the debts incurred by its predecessors, tried to balance the books by making expenditure cuts, all government departments suffered savage budget reductions. As part of this measure, public servants took a 10 per cent wage cut.

By 1933 the Log Pool was still providing excellent sport, with two rods reportedly taking 18 fish in three hours one morning;[23] and this, one week after one of the largest floods for many years. But the threat of erosion remained. The concerns of Tongariro regulars made them put their rods down for half a day at the start of the 1934 season and hold a meeting with the government engineer, L. May, to examine what might be done. May advised planting willows to secure the true left bank, and using a groyne at the tail of the Hut Pool to divert the water flow in a direction that would maintain the Log and other pools downstream.[24]

Over the next few years, the erosion seems to have been a continual threat. Access to the fishable water required wading a dangerous channel, but the Log Pool remained a favourite of the Tongariro regulars. The trout must have liked it as well, for Frank Reynolds, a Whakatāne angler with much experience on the river, took 12 fish from the pool early one morning at the start of the 1936 season.[25] One of the real old hands on the river, J. S. W. Neilson, banked a limit in early March 1938, one fish being a rainbow of 10lb.[26]

The erosion problem remained for the next few years as different government departments argued over what should be done and which one of them should carry the costs of the remedy. The Department of Internal Affairs wanted the Native Lands Department to contribute, because it was thought clearing vegetation from the banks on the lower river had made erosion worse. In the end, it was decided not to install groynes in the riverbed, but instead to cut a channel at the tail of the pool above so as to turn the course of the river. This method lowered the level of the water feeding the pool and resulted in better fishing.[27]

Despite the anglers' fears that the Log Pool would be lost, it continued to provide superb trout fishing into the 1940s. On one day in April 1941, two 12lb trout were caught.[28]

The next threat to the pool came with the great flood of 1958. After the river had settled, an attractive lie remained, but the fish never seemed to hold there in the same numbers.[29] The anglers moved on to other, more promising pools, such as the Major Jones, which had kept its reputation intact.

Dreadnought Pool

The Dreadnought Pool is another lost to anglers after the great flood of 1958. The pool gained its name from the mighty bluff which commanded the river as it united at the base of an island dividing the water further upstream. This bluff, resembling the prow of a warship, jutted headlong into the plunging rapids to form a wide, wadeable pool which could accommodate several rods.

It is not known when the name Dreadnought was first applied to this stretch of water. England built her battleship *Dreadnought* in December 1906 in a mood of jingoism which started the naval arms race with Germany. Fly fishing began in earnest on the upper Tongariro River after 1910, so at a guess the Dreadnought Pool could have been named before World War I began.

The pool was first referred to in the New Zealand press in April 1927, when *The New Zealand Herald* mentioned Romer Grey, son of Zane Grey, catching a 15¼lb rainbow trout there.[30] But while Zane Grey may have made the pool more widely known by writing about his experiences there, it is certain he did not name it.

Although Zane Grey was eventually enchanted by the Dreadnought Pool and caught some of his best trout there, it took some time for him to fall under its spell. In 1926 it took real determination to get down to the water where you could cast a line. First, Māori guide Hoka Downs took the party through a narrow, fern-fringed track to an apparent deadend at the top of a small bluff. The only way down was to jump with tackle in hand and hope for a soft landing. All Grey's group managed it without injury, including Morton, the photographer accompanying them with his cameras and heavy boots.

View of an unidentified man fly fishing at Dreadnought Pool, Tongariro River, Waikato Region.

Ref: WA-04362-G. Alexander Turnbull Library, Wellington, New Zealand. /records/30649339

After that it was a scramble across a narrow plank bridge supported by wires, and then more fern bashing, before they reached the riverside. Hoka Downs gave them a brief tribal history; it was here that his people had fought battles and maintained a pā on the top of the bluff.

Grey opened his account with a superb 11½lb male rainbow, which he did not kill immediately but kept in the river, tethered on a cord where he could admire it. More trout followed, but for his fellow angler Captain Mitchell it was an unlucky day when he broke his rod. Although Grey wanted them to take turns fishing with his rod, Mitchell insisted that Grey should fish on and take a good bag. Grey surrendered his rod only when the cold water forced him to stop, but even then Mitchell raised no trout.

Zane Grey fished the Dreadnought Pool four times in all during his first visit in 1926, devoting two chapters to it in his book *Tales of the Angler's El Dorado*.

Another writer who fished the pool some 25 years later was Geoffrey Hobbs, author of *Fisherman's Country*. He describes how he and his companion copied other anglers and cast lures across the rapids at the head before fishing the lower reach carefully, but met with no success. Even the greatest pools could sometimes disappoint the angler.

Today, the Major Jones Pool is one of the few stretches of river that remains from the very early days. For the Hut, Log and Dreadnought pools, we are left with only memories and archived photographs.

CHAPTER 16:

World War II years and later

Research is today the most essential tool in the hands of the management authority. Gone are the days when a fishery such as Taupo could be managed by hearsay.

— PAT BURSTALL, SENIOR FIELD OFFICER — FISHERIES, DEPARTMENT OF INTERNAL AFFAIRS, FEBRUARY 1959[1]

ON 3 SEPTEMBER 1939 NEW ZEALAND joined Great Britain in the fight against Hitler's Germany. The commitment to the Old Country was as natural as night following day. Like Australia, New Zealand was a self-governing Dominion but still felt the bonds of Empire and Commonwealth. For the next six years events on the global stage occupied the nation's leaders as they poured men, materials and food into the war effort.

Wartime travel and supply restrictions

Travel restrictions became the order of the day for most people, although not immediately. The government imposed petrol rationing early on, the limit being 36–54 litres per month, depending on the size of the motor car. This would later seem a generous amount, as in 1942 the maximum was 9 litres per month for private motorists.

With rubber supplies under threat, car tyres were also in short supply, further limiting the opportunity for private travel.

In December 1941, the government announced without warning that rail travel would be restricted in the forthcoming holiday period, which was traditionally a popular time for angling in the thermal regions. A blanket rule prohibited excursion train journeys of 100 miles or more unless there were special reasons. Despite these difficulties in getting to Taupō, some anglers from the nearby town of Taumarunui managed the trip to Tokaanu by road.[2] The large trout were still there for the taking.

Nevertheless, as New Zealanders adjusted to the restrictions, the anglers on the Tongariro became fewer in number. The conflict also prevented the normal influx of overseas visitors, which badly affected the fishing launches operating on the lake. Frederick W. Pickard, a United States angler, fished the Tongariro River for just one visit, in November 1939, two years before the United States entered the war. He described fishing the Dreadnought Pool with a dry fly in the company of another angler[3] with some success.

By the start of 1942 *The New Zealand Herald* reported that Taupō had few angling visitors because of the emergency regulations.

Although the flow of visitors slowed, it did not cease entirely. Taylor's Lodge entertained the Governor of Fiji in March 1943.[4] The armed forces, who had greater freedom of movement, also found time to sample Tongariro trout. The licence fees were altered to allow any member of the Army, Navy or Air Force and other visiting service personnel not resident in New Zealand to buy a fishing licence at a reduced cost. Lodge bookings for the start of the 1943–44 season were good.[5]

Thanks to these changes, some United States marines had time to experience the thrill of landing Taupō trout before heading off to face the Japanese troops in the Pacific battlegrounds.

The weight of trout caught remained high, with double-figure fish appearing at times. During 1943–44 one of 14lb was taken, while next year was even better. The official report for 1944–45 noted 40 large trout caught, with weights ranging from 10lb to 18½lb. These results may have stemmed from the presence of smelt in the fishery. This small bait fish had been introduced from the Rotorua lakes in the 1930s, and by 1941 were said to be established all over Lake Taupō 'in countless thousands'.[6]

The Department of Internal Affairs also transferred some trout within the thermal district, perhaps to broaden the species' gene pool. In 1943 fry raised from ova taken from Lake Tarawera trout were released in the Tongariro and Waitahanui rivers, while Waitahanui River fry were sent to a new home in lakes near Rotorua.[7]

Wartime conditions also meant people had to take extra good care of their fishing tackle, because new equipment was not easily obtained. In 1944 a newspaper noted that after five years of war very few shops had fly rods for sale, although stocks of split cane were available to deal with rod repairs. Waders had to be patched and re-patched, only secondhand fly lines could be bought, and hooks were in short supply.[8] An unexpected effect of wartime conditions was the difficulty anglers had in obtaining silkworm gut for leaders, as the gut was reserved for surgical use.[9] Petrol restrictions were removed several years after the war's end, but it took some years before anglers could buy imported trout tackle freely.

Lake Taupō raised

In 1941 the government installed control gates at the Waikato River outlet to raise the level of Lake Taupō several feet (about 1 foot below the maximum level). The lake was at its highest in spring, but peak power demand came in May and June. Storing spring rainfall would mean more efficient power production at the Arapuni power station.

A number of lakeside properties were affected, including those at Tokaanu, which was already a low-lying site. Sewerage systems at the Tokaanu Hotel were rendered useless,[10] and Delta Camp flooded. After the authorities had kept the lake at a high level for three years, it became clear that some trout feeding areas had been lost but new ones added. Overall the lake scheme had had a detrimental effect on the fishery, especially on the eastern side. However, the presence of smelt kept fish in good condition.

Fisheries management reviewed

As we saw in earlier chapters, the best way to manage the Taupō fishery, so as to ensure quality fishing for the future, became a vexed question for government officials. From time to time the press would report there had been a disappointing season, with the trout taken being smaller than the year before, in poorer condition, or both. Of course, even the

Effect of raising the lake on Delta Camp.
Advocates for the Tongariro River, from the collection of Mrs E.G. Renz

greatest fishery will give fluctuating results from year to year, but every poor season noted in the press provoked adverse comments.

Anglers as a body are great complainers, and when the Hon. W. E. Parry, Minister of Internal Affairs, visited the thermal lakes on an inspection tour in 1936, he received plenty of free advice on how to improve matters — additional fisheries experts, raising or lowering of lake levels, the introduction of more fish food, better riverbank access tracks, and increased stocking of waters. Although these were the opinions of experienced anglers, they were little more than idle speculation. The only conclusion Parry took away from his tour was that no one could agree on what the problem was, let alone a solution.[11]

At the end of World War II, officials at the Department of Internal Affairs realised that action was needed to deal with problems affecting the nation's flora and fauna. These problems included declining numbers of native birds, soil erosion caused by deer and other browsing animals, and an apparent decline in New Zealand's trout fisheries. The Wildlife Branch was set up within the department to tackle these problems. A new Field Investigation Unit would provide a link between the staff who observed what was occurring in the outdoors and the scientists formulating wildlife management policy.[12] Several staff members were seconded to the Marine Department to undertake training in freshwater fisheries operations.

Derisley Hobbs investigates

By now Taupō was the most popular fishing destination in the thermal fisheries region: district fishing licence sales for Taupō in 1947–48 were 8,801, while those for the Rotorua district numbered 6,251. Towards the end of the 1949–50 fishing season, Taupō anglers became concerned over the gradual reduction in the size of the trout they were catching, compared with previous years. Archie Kitto of Whanganui wrote an angry letter to the press complaining that the government 'turns a deaf ear' to anglers' pleas for improvements to the fishery. The 1949 season had been a dismal failure, with catches averaging only one fish per day. His suggested remedies included blasting the waterfall at Beggs Pool on the upper river to remove it as a barrier to trout, netting the lake, and extending the closing date for the fishing season.[13] It appears this agitation may have had an effect, for after Taupō anglers made submissions to the Minister of Internal Affairs, the Hon. W. A. Bodkin, the minister instructed the Marine Department to investigate and report. This job fell to Derisley Hobbs, the Senior Fisheries Officer and brother of G. B. Hobbs, author of the fishing book *Fisherman's Country*.

A second concern for anglers was the visible deterioration in many of their favourite pools in the Tongariro's lower reaches. A series of eruptions by Mt Ruapehu in 1945 had deposited volcanic ash in the watershed, so that what was once good holding water in the Cherry and Poplar pools had shallowed and no longer provided a haven for migratory trout.

Hobbs visited the Tongariro in May 1950, and studied the situation to see if the matters raised had substance and could be remedied. He started by visiting the upper river, beginning at Beggs Pool, and worked his way down to the delta.

From de Lautours Pool to the mouth, he found the riverbed to be composed of coarse black sand with some pumice deposits. This sand was a constantly drifting presence, even at times of low water flow, and its abrasive action on the bed of the river meant no algal matter could gain a hold, thus rendering the food chain non-existent. The problem had been accentuated by the raising of the level of Lake Taupō in 1941. A higher lake meant that the Tongariro's lower reaches had a lower gradient. The effect was that silt which would normally have reached the lake remained on the riverbed, where only a major flood could shift it. Some cutting of channels might give the river a new course to the lake, and, if dredging and stopbank construction were carried out, a solution might be found, but it would probably be of a temporary nature only.

Hobbs quickly realised that not enough was known about the Taupō fishery to allow a clear picture to be drawn. Even though fish size and condition had been measured season by season since the 1930s, resulting in much data, it was selective and poorly balanced. One major omission was a survey of the size and condition of fish outside the November to May fishing season. However, by collating all the data recorded since the beginning of the fishery Hobbs could see where there were gaps in the information and plan to fill them. It was immediately obvious that poorly conditioned trout were always present to some extent, even in the fabulous years of the early 1920s. Secondly, the proportion of undesirable fish increased sharply in proportion to the size of the fish being studied.

To allow a proper survey of the fishery to be completed, the Marine Department negotiated with angling interests and reached an agreement that no changes should be made to the Taupō fishing regulations for at least three years. The objective was to remove any unnecessary complexity to the extent possible while the research programme was in progress.

Pat Burstall arrives

An important change occurring in the 1950–51 season was the arrival in Tūrangi of a young fisheries officer employed by the Wildlife Branch of Internal Affairs — Pat Burstall. Burstall had completed his training as a fisheries technician, and began operating at Tūrangi collecting fisheries data for the scientists of the Marine Department. He made his mark immediately by getting more anglers to keep diaries of their catches over the season and hand them in to the department. When the position of ranger became vacant at the end of 1951 he was appointed to that role, which was to be performed as part of his fishery survey duties. His superiors saw immediately that he had a keen interest in fisheries biology, and the function of ranger (which came with a vehicle) was a way of boosting his status. But Burstall's enthusiasm was to cause friction with his superiors almost immediately.

Derisley Hobbs suggested a wide range of duties for Burstall under the research programme:

- collecting diary data from across the Taupō region
- conducting a summer survey of potential spawning streams, starting with the rivers located on the western side of Lake Taupō
- repeating the stream survey in winter and obtaining comparative data on fish numbers
- taking detailed samples of fish from spawning streams
- collating data taken and submitting concise monthly reports to the Marine Department.

It soon became obvious that fisheries data was accumulating faster than Pat Burstall could deal with it. He and his wife, Robyn, worked at night typing reports (using their own typewriter), but the task was hopeless. In April 1952 Hobbs recommended another fisheries officer should take over the job of collecting data for the Marine Department, although Burstall would still complete the monthly summaries. That step was taken, and by late August Burstall was ordered to hand fish trapping returns and anglers' diaries for 1951 to another officer. It later became clear that Burstall passed on only a part of the information. When the Conservator of Wildlife, Stuart MacNamara, pressed Burstall on the point he undertook to send the remaining information, but followed up with written advice that he wished to leave the department in late November 1952, some three months hence.

It seems Burstall's intense personal interest in fisheries research made him think he was the person best suited to process the raw data, even though this was purely clerical work which fell outside his responsibilities.[14] He also seems to have feared that if he did not tabulate the data collected immediately, the enormous amount of surveying in the field would be wasted. That gave him a proprietorial attitude to the work.[15] The picture of Pat Burstall that emerges is one of an intense, committed individual with his own ideas of how the fisheries research should be carried out.[16]

Frank Yerex, the Controller of Wildlife, smoothed over the contretemps by suggesting Burstall should prepare several copies of the monthly summaries and send them simultaneously to Hobbs and the Wildlife Department, with Burstall keeping copies for his own records. Yerex also stressed that Burstall and other fisheries officers would be given recognition for the part they played in the research.

It is clear that Hobbs was quite prepared to allow for enthusiasm and diligence on the part of fisheries officers collecting data in the field, but he drew the line at them having a role in interpreting that information. That was a job for the scientists. But while Hobbs felt the 'data arising from the research programme is not [Burstall's] to analyse or interpret', Burstall wanted to partly share in the final presentation of the work. Burstall didn't help his case by embarking on a history of the Taupō fishery. In mid-1953 Hobbs recommended that Burstall should be relieved of responsibility for fisheries work in the Taupō area; a move that Burstall considered unjustified and ungrateful.

The 1953 survey results

Over the winter of 1951 the Department of Internal Affairs fisheries team ran fish traps on tributary streams, measuring and marking some 6,000 trout. This survey gave the first comprehensive picture of the composition of the larger trout in the Taupō fishery. The process was repeated in 1952. By November of that year Hobbs was able to predict that an answer to the problem of poorly conditioned trout would be found.

The official report released in 1953 concluded that the Taupō trout stocks were fundamentally sound, but the fish population contained an excess of older fish that had failed to regain condition after spawning. Selective fishing by anglers (rejecting poorer-conditioned fish) had made things worse.

Secondly, while the fishery's annual yield of 200,000 fish (equivalent to 1 million lbs) was greater than before, there was scope for an increased yield if anglers killed a good proportion of the poorly conditioned fish instead of cherry-picking the best ones. That approach would divert the available food from 'unthrifty' old fish to strongly growing younger ones and thereby increase the average size.

As a short-term measure, officials recommended extending the fishing season and increasing the bag limit from six to eight fish, with all trout landed to be counted. But the truly adventurous proposal, intended to take the investigation to a higher level, was to have a supplementary fishing season in the winter of 1954 (1 June to 31 October). This was seen as a research experiment which carried little risk; if too many trout were taken then the average weight would increase, while 'the probable gain will be knowledge which will settle once and for all the questions which have worried anglers and the Department for many years — namely, what is the ideal duration of the angling season for Lake Taupo, and what adjustments can be made to ensure that the highest proportion of fish feasible will be taken when in good condition'.[17]

The experimental winter seasons

A winter season running right through to the start of the next open season was a radical move which went beyond anything the Taupō angling clubs had been advocating. The fisheries scientists saw it as an experiment depending on close observation of both fishers and fish, and in this respect it was successful; all the data that could reasonably have been expected was obtained.[18]

Conclusions drawn from the data included:

- the experimental season saw an estimated 36,500 trout taken by anglers, which was about one-fifth of the harvest in a normal seven-month open season
- angler effort expended was around one-quarter of that for an open season
- most angler effort was directed at mature fish in the Taupō rivers — the reverse of the summer angling trend
- mature fish in the rivers averaged 22½ inches in length and 5lb in weight
- the proportion of poorly conditioned fish rejected by anglers during the winter was 5 per cent for the lake and 4 per cent for the rivers
- the rate of catch per hour in the rivers did not change markedly in the winter period, while in the lake it was very low but gradually improved.

Hobbs and Yerex, of the Marine Department and Department of Internal Affairs respectively, agreed that the main problem with the Taupō fishery was the high proportion of recently spawned, out-of-condition fish which were caught in the summer period when most anglers preferred to travel to the Taupō region. They largely solved this problem by recommending that anglers lower their standard and regard trout in fair condition as acceptable, where previously anglers had tended to reject any fish not in prime condition.

In early February 1955 the officials decided that another winter season (June–October) should be allowed that year for research purposes, with the bag limit raised from five fish per day to eight.

For the 1955–56 open season they proposed adding October and June.[19] Provision for annual licences (1 July to 30 June) was made by regulations issued in 1960.

The second winter season gave results similar to the first one, although fewer fish were caught (33,600 in 1955, as against 42,000 the previous year). A notable feature was the consistent length of Taupō rainbows from 1951–52 to 1954–55 (an average of 22.0–21.8 inches). By the end of 1955, Derisley Hobbs was satisfied that the project's principal objectives had been achieved, one of which was to eliminate unanswered questions about the fishery. As a result, there was a better picture of the Taupō fishery than in any other lake fishery elsewhere in the Southern Hemisphere. There had also been a reduction in the number of 'nuisance' fish not wanted by anglers. Hobbs praised Marine Department and Department of Internal Affairs staff for their teamwork, and noted the co-operation by anglers in the research programme. Despite these advances, many matters remained for investigation, including the factors affecting the seasonal abundance of food organisms.[20]

The Tongariro River played a significant part in the research work. Details of Tongariro fish caught were compared with those passing through the fish trap in the Whareroa Stream, a tributary on Lake Taupo's western shore.

Because of the fishery's history, Hobbs thought a larger bag limit and a longer open season were essential. He was anxious to avoid a repeat of the boom-and-bust days which had occurred some 40 years before: if the fish stocks in the lake were reduced sufficiently, the average size of the fish at their first maturity was sure to increase.

By November 1954, after the first experimental season, Pat Burstall had formed his own theory. He concluded that there had been a build-up of inferior trout in the Lake Taupō population since the mid-1920s. The introduction of smelt as a food fish in the 1930s had halted the decline temporarily, and it was not until 1950 that anglers complained about the declining quality of their catches. From spawning records, Burstall concluded that anglers should be able to fish the Taupō rivers in August, when the late-running mature fish were present. Progressively culling them over the years would divert food to younger trout and reduce the problem of superimposition — a process whereby eggs laid by trout in the early part of the spawning season were disturbed and overlaid by eggs of trout which spawned later on.

Personal and interdepartmental conflict

In late 1955 Burstall was appointed Senior Fisheries Officer for the Department of Internal Affairs at its Rotorua office. He was now responsible for fisheries management in the Rotorua and Taupō regions. Hobbs gave up his job as Marine Department Senior Fisheries Officer to take on a broader administrative role, but continued to have some influence in a consulting function in relation to freshwater fisheries.[21] The two men had had a difficult working relationship, even though Hobbs had given the younger man some leeway from time to time to pursue research topics of purely personal interest. The friction was largely caused by Burstall's unwillingness to accept proposed management strategies for the Taupō fishery, but it also occurred against a background of interdepartmental conflict at the time. The fundamental issue concerned who was best placed to decide management strategies: the fish and game staff employed by the Wildlife Branch of Internal Affairs, or the Marine Department's scientists? The argument began shortly after World War II and continued into the 1960s.[22]

In 1959 the Controller of Wildlife asked Burstall to prepare a report on the state of the Taupō fishery. Burstall repeated his claim that, based on the evidence collected so far, the fishery was carrying too many old fish which were biologically 'uneconomic'. These fish consumed food items that would otherwise have been available for younger trout. As a result, the rivers could bear greater angling pressure during the spawning season, provided the spawning areas were protected. Rather surprisingly, he felt that superimposition of ova and poaching had been the major limiting factors on the rainbow trout population, with angling pressure playing a lesser role.

Pat Burstall survived the political ructions, and several years later went on to play a major role in safeguarding the Tongariro trout fishery from the worst effects of the power generation scheme that was to come. Without his practical experience and detailed knowledge, far more damage would have been done.

CHAPTER 17:

The great flood of 1958

Turangi was completely isolated. To the north the bridge was breached; to the south all bridges on the Desert Road were down; to the west Tokaanu was floodbound with a couple of slips . . .

— HAROLD HICKLING, *FRESHWATER ADMIRAL — FISHING THE TONGARIRO RIVER AND LAKE TAUPO*

FOR TONGARIRO ANGLERS THE 1950S were to end with a cataclysmic event in the form of a massive summer flood. High water was nothing new for the river, as large floods had inundated Tokaanu in mid-January 1907,[1] damaging the main highway bridge. There were also major spates in 1924, 1933[2] and 1935. By coincidence the 1935 event also occurred in late February following a prolonged dry spell.[3]

The scope of the 1958 flood

What made the 1958 event notable was not merely the size of the flood — the highest ever recorded — but the permanent alterations to the fishing pools and the damage to Tūrangi township.

The event began with a relatively small flood (170 cumecs) on 18 February. From 17–19 February, 5.1 inches (almost 13 centimetres) of rain fell, but even that exceeded the normal rainfall for the whole of that month. No rain fell on 20 February, and the river level fell back to normal. Heavy rain then began again and continued for several days, with the waters rising again on 22 February. The Tongariro flood peaked at 10am on 24 February, when the flow reached an estimated 51,000 cusecs (cubic feet per second; 1,444 cumecs [cubic metres per second]). Total rainfall for the month was measured at 22.44 inches (57 centimetres).[4]

The speed at which the river rose — 15 feet (4.5 metres) in one hour, according to eyewitness Vice-Admiral Harold Hickling, RN (Rtd) — seemed mysterious. Hickling speculated that trees and other debris became jammed in the narrow gorge section in the upper reaches, causing the river to form a temporary dam which eventually burst, unleashing a massive torrent. Yet when Pat Burstall examined the rainfall records for the district, he found all of the Taupō tributaries had risen in equal proportion.

As for damage the flood caused, the report made by Department of Internal Affairs staff once the floodwaters had receded resembled a battalion roll call after an attack on the Somme: the casualties were heavy and some familiar names were missing. The footbridge over the Red Hut Pool suffered severe damage, with one of its massive concrete supports being shifted out of position. The anglers' bridge spanning the Major Jones Pool fared better, losing only its central decking, and so could be repaired promptly.

At the hatchery camp, now the site of the Tongariro National Trout Centre, the floodwaters rose to a depth of 3 feet, and on receding left a layer of silt 6 inches deep.

Mere statistics cannot possibly indicate the potency of the river's fury. The best personal account remains that of Hickling, an angler and Tūrangi resident who had the river at the bottom of his garden. In his book *Freshwater Admiral* he devotes a chapter to the effects of the flood:

> I have seen many rivers in flood, the Yangtse, the Potomac in 1943 when only levees of sandbags kept the waters at bay from the White House, the Hoogli; the Tongariro at that moment was something I had never seen before.
>
> True the river was only a hundred yards wide but the force and speed of it was terrifying. It was running at a good twenty knots and you couldn't hear yourself speak for the roar of the turbulent brown water which was then lapping over the top of the bank, some twelve feet above the normal level, . . .[5]

The 1958 flood damaged both river and houses.
W. G. Henderson

A convoy of guests staying at the Bridge Lodge vacated smartly, intending to drive north to Taupō town. After crossing the Tongariro River bridge at Tūrangi, their progress was halted several kilometres later by a smaller stream that had overflowed its banks. Returning to the Tongariro, they found that a large part of the bridge there was partly destroyed, marooning them for four days.

Harold Hickling helped his neighbours to rescue animals and property before mourning his own loss. As we saw in Chapter 11, he had spent some months writing a book on how to fish the Tongariro River, complete with detailed diagrams of each pool and the best lies. Overnight, his carefully drafted manuscript (originally called 'The Tongariro Tale') became redundant and he had to start afresh. The result was *Freshwater Admiral — Fishing the Tongariro River and Lake Taupo*, published in 1960. The book pre-dates the rise of the weighted nymph as a technique, focusing instead on the downstream wet fly, but its combination of angling history, instruction and anecdote made it a valuable addition to New Zealand's fly-fishing library.

New pools for old

In assessing the post-flood situation, Pat Burstall noted that some existing pools had been destroyed but new ones created. Overall, he felt the value of the fishery had not been seriously jeopardised.[6]

On the upper river the greatest loss was the Dreadnought Pool, popularised in the 1920s by Zane Grey and his party. The Fan Pool immediately above it — superb holding water often favoured by Harold Hickling, Archie Kitto and other Tongariro regulars — had turned into several shallow rapids.

Further downstream, below the road bridge, several smaller pools — such as the Swirl, the Nursery and the Stones — were no more, as was the famous Hut Pool, although new fishing water resulted, fishable from both banks. The Hut Camp buildings ceased to exist.

The Hut Pool gave superb sport right up to the eve of the flood. In a delightful cameo, a fly fisher equipped only with an 8-foot rod fished the Tongariro's fabled water using a No.10 Twilight Beauty and a leader with 5lb breaking strain. He hooked a fish in the rapids at the head of the pool and, in a reversal of the usual procedure, the fish played him for an hour until he ran it ashore because his small net would not fit its 8lb bulk. That night heavy rain began, and within 24 hours the Hut Pool was no more, but the final trout it yielded was a memorable one.[7]

But Nature can give as well as take, and a post-flood survey found that the alterations to the riverbed had bestowed some benefits. The Birch Pool was greatly improved when it changed into two pools (Birch and Upper Birch), while the Bridge Pool became one of the most productive lies on the river. Further down, the Boat Pool was enlarged, while the Log Pool, the Reed Pool and the Jones Pool were found to be in better condition than before.[8]

Once the waters receded, Tūrangi residents could see the effect on their properties. Eighteen riverside cottages had been swept away, while half the remaining houses were damaged, with inches of river silt coating their floors. The flood severed a portion of the main highway bridge, which meant Tūrangi was isolated for several days. The New Zealand Army came to the town's rescue by erecting a prefabricated truss bridge.

The 1958 flood was a one-in-100-years event; the river god had shrugged. As the seasons passed, trout returned to the river and the fishery re-established itself. But by 1960 plans were already afoot that would see the river changed forever.

CHAPTER 18:

The Tongariro power development scheme

500 Men Poised for 'Plumbing Job'
— ***THE NEW ZEALAND HERALD*, 18 JULY 1964**

ON A WINTER'S DAY in early August 1964, 35 men sat in a room in downtown Wellington listening to a government official, J. T. Gilkison of the Ministry of Works and Development. Gilkison, along with staff from the Marine Department and the Department of Internal Affairs (Wildlife Branch), was doing his best to persuade those present that a massive government energy project, the Tongariro hydro-electricity scheme, was essential to New Zealand's energy needs.

But Gilkison had a problem; many in the group weren't buying his message. As representatives of the Lake Rotoaira Trust Board, various acclimatisation societies and Taupō trout fishing clubs, they wanted to see the Tongariro River and other precious trout streams preserved, or at least protected as far as possible. Gilkison asked the delegates three times to support the project, but the results were at best mixed. While no angry words were spoken, reactions after the meeting had ended ranged from reluctant acceptance to strong opposition.

'I'm not happy with assurances,' said John Henderson of the Wellington Acclimatisation Society. 'I think there will be regrets.'

Similarly doubtful was P. J. Hura of the Rotoaira Trust Board: 'From what has been said, it seems to me it depends on trial and error.'[1]

When inviting the acclimatisation societies and clubs to the meeting, the Department of Internal Affairs had shown considerable trust. It disclosed the Marine Department's views about fisheries aspects of the scheme and asked for these to be kept confidential. The trust seems to have been respected.

Power planning

The Tongariro power development scheme was authorised under the Public Works Act 1928 — a very powerful piece of legislation. With the Town and Country Planning Act 1967 still several years away, New Zealand lacked any formal procedures for scrutinising major projects from the standpoint of their effects on the physical environment. The lack of formalised, open procedures for evaluating the Tongariro power development scheme (let alone challenging it), and the ridiculously short timeframe in which the government attempted to operate, were the main causes of the public's subsequent anger.

Energy generation was an important issue for the New Zealand government in the 1950s. Officials had conceived the idea of taking water from rivers flowing off the central North Island plateau, impounding it and then diverting it into Lake Taupō. Because Lake Taupō was the source of the Waikato River, the additional water would boost the output of the power stations downstream, having already been used for power generation in the Tongariro River catchment before it reached the lake. The scheme was attractive to the government because it could be implemented in stages and modified as construction progressed; a sort of trial-and-error approach. There would also be little call on 'overseas funds'; that is, funds produced by the nation's exports and held in overseas currencies by the government, this being a time when New Zealand operated a closed economy with a controlled exchange rate.

For months, angling groups and other concerned citizens had complained that the scheme was shrouded in mystery and not enough was known about the possible environmental effects. The government had called the Wellington meeting to curb opposition by providing information on how the project would operate and the supposed benefits to the trout fishery.

A British engineering firm, Sir Alexander Gibb and Partners, was asked to advise on the merits of the proposal. Its brief required that allowance should be made for the preservation of fisheries and scenic

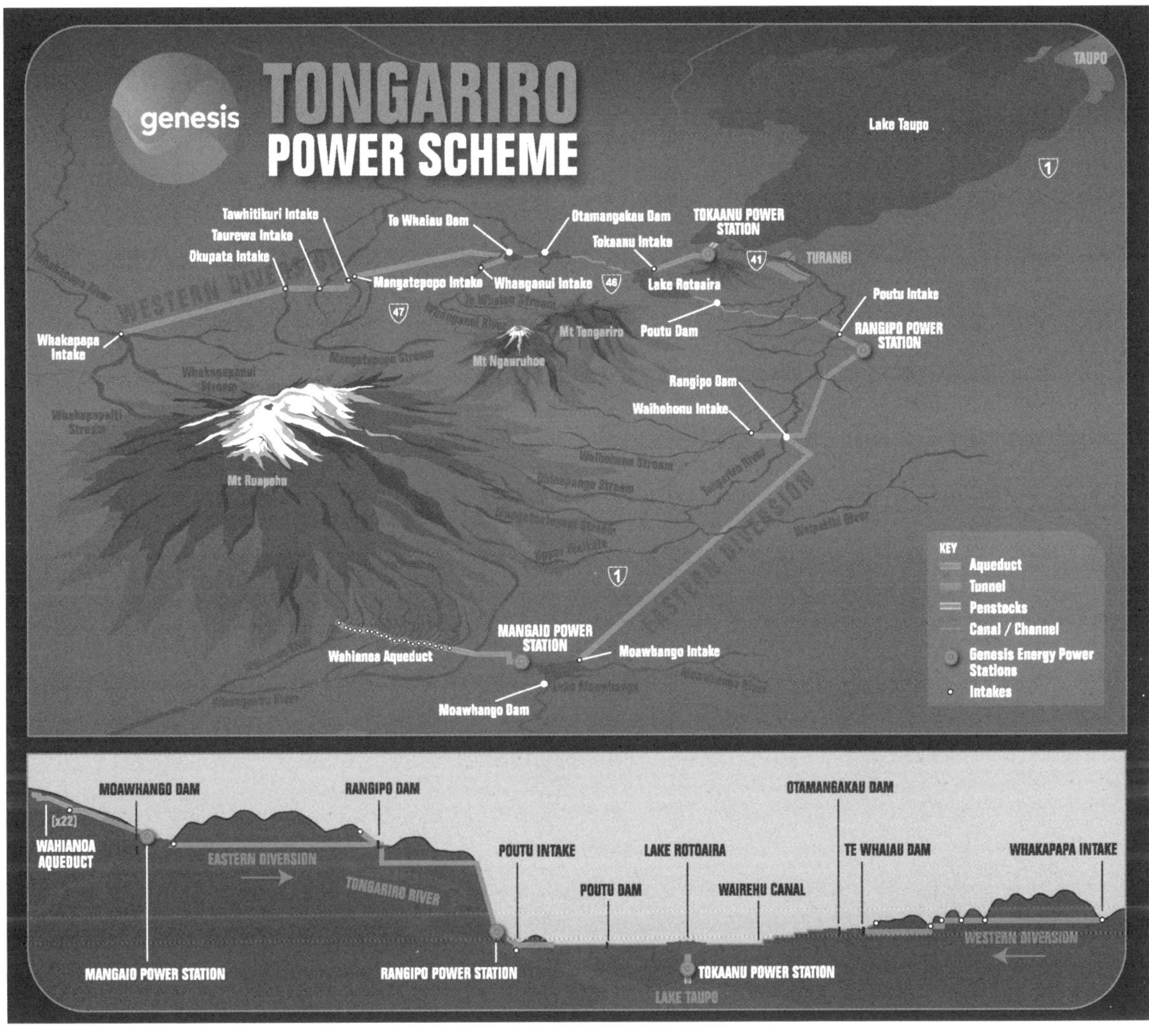

The Tongariro power development scheme involved many kilometres of tunnels linking a series of dams and generation stations.

assets. The Marine Department and the Department of Internal Affairs (Wildlife Branch) were also brought in to give their views.

In October 1958, a Marine Department confidential memo set out the department's position:

> Key principles
>
> . . .
>
> 1. 'The Tongariro river is an integral part of the Taupo fishery, which is the greatest trout fishery in the southern hemisphere.
> 2. The striking increase in the demand for electricity is not, in terms of growth, any more remarkable than the increase in the demand for angling.

3. It is imperative that the new Hydro-electric undertaking should be integrated with, rather than developed at the expense of freshwater fisheries. To the maximum extent feasible, the two needs must be reconciled.'[2]

At first, anglers and other recreational groups in the Taupō area knew that the Tongariro was threatened, but had little information to go on. In May 1957, a conference attended by Department of Internal Affairs officers, members of the Tongariro and Lake Taupō Anglers Club (TALTAC), and other angling and boating groups passed a remit saying that the harnessing of the Tongariro River for hydro-electric purposes was strongly opposed. The meeting was informed that the Controller of Wildlife (now Henry Kelly, who had replaced Frank Yerex) was a member of an interdepartmental committee formed to consider the effects of the power scheme, and he would protect anglers' interests as far as he could.[3]

That statement accurately encapsulated the position of the Department of Internal Affairs' Wildlife Branch and the Marine Department. As public servants with an advisory function, they were bound to implement the government's policies. All they could do was uphold the interests of their respective departments by preventing any adverse effects arising from the scheme and reducing the impact of those changes which could not be avoided. No formal opposition to hydro-electric development could be mounted unless it came on a political level from the public.

The surveys for the project went on. The activity was obvious to the people of Tūrangi and Taumarunui, who remained watchful but were confronted with a threat that had not fully crystallised. TALTAC had already objected to the proposal to divert toxic Whangaehu River water into the Tongariro.[4] The club protested to the Department of Internal Affairs about the threat to the Tongariro fishery, but was not satisfied with the vague assurances received in reply. In October 1960, members voted to inform the government that the Tongariro diversion was unacceptable to Taupō anglers.[5]

The Woods report

The Marine Department then gave one of its freshwater biologists, Cedric Woods, the enormous task of evaluating the effects of the Tongariro power development (TPD) on the Tongariro River and other trout streams flowing off the mountains of the central plateau region. Woods used previous fisheries surveys to analyse the hydrology of the area, the likely effects (adverse and beneficial) of the project, and what those would mean for trout stocks. He was careful to note that he was dealing with major fisheries problems that had to be answered before the plan could be finalised, rather than those which might arise in the course of construction.

After the consultants gave a favourable report, in 1963 the government's Power Planning Committee recommended the scheme should go ahead, even though Woods had only formed a draft set of conclusions. Anxious to get on with matters, the government approved the Tongariro power development scheme on 23 March 1964 — albeit only in principle. The official justification was a combined Electricity Department and Ministry of Works report stating that the scheme was the cheapest option remaining in the North Island for power development. Approval was given 'subject to [the government] being satisfied that suitable arrangements can be made to preserve the interests of parties who would be adversely affected by the scheme'.[6]

Around that time Woods completed his survey for the Marine Department. The report was the result of three years' work and ran to more than 200 pages. Its main thrust was that while the flow characteristics of the Tongariro River would be altered, the trout fisheries in general would not be adversely affected and the power project could be carried out without material damage to the river.

Woods thought that the power scheme could maintain the quality of the Tongariro fishery, and possibly even improve it. While the flow in the river from the Poutu diversion down to the delta would be reduced and the river's channel would be narrowed, the proposed guaranteed minimum flow of 1,000 cusecs (28.3 cumecs) at Tūrangi would mean the fishing pools would be preserved.

One of his main recommendations was for a pattern of artificial freshes (floods) to supplement the Tongariro River's natural flow at specified times in the season (in November, December and from January to May). That would bring trout up-river just when anglers needed them most — at holiday times, for instance. '"Made to Order" Sport Possible in the Tongariro' trumpeted *The New Zealand Herald*, when it reported on Woods's findings.[7]

From the government's standpoint, there was an irreconcilable conflict between the need for an increased hydro-electric power supply and the preservation of scenery, including rivers and trout fisheries. If the nation's industries and infrastructure were to expand for the greater economic good, then the power would have to be generated somehow. Whatever political opposition was aroused it would almost certainly have to be overridden, because no sources of power other than the central North Island rivers could be in production by the projected 1970 commencement date.

Anglers react to the scheme

To some in the angling community, the Tongariro power development represented a massive threat to the world's most famous trout fishery. Con Voss, fly casting champion and president of the Rotorua Fishing and Casting Club, considered that any interference with the natural flow of the Tongariro River or other Taupō waters was bound to have detrimental effects. To spoil the fishing in what was the foremost fishing river in New Zealand, one with an international reputation, would be a tragedy. Voss also pointed out the enormous damage that would be done to the Whanganui River if its feeder stream, the Whakapapa, was diverted; a point many others had failed to notice.

Hud Little, president of the Tongariro and Lake Taupo Anglers Club (TALTAC) was even more pessimistic. He predicted the entire scheme would lead to the total ruin of the Taupō district as a fishing and tourist attraction. TALTAC's protests, he said, had been ignored by the government, which was apparently intent on bulldozing the scheme through, despite public opposition.

Another opponent was A. V. 'Bruno' Kemball, who sold beautifully tied flies to anglers from his small shop tucked away among the trees at Hatepe, on Lake Taupō's eastern shore. To him, the Tongariro power development was the brainchild of a boffin, created without any thought for the spawning beds in the river. 'The Tongariro is the finest fishing river in the world,' he said. 'And anyone who spoils it deserves to be horsewhipped.'[8]

Others felt the scheme might have some advantages. Geoff Sanderson, the Tūrangi fishing tackle retailer (and president of the Tokaanu–Turangi Chamber of Commerce) thought the benefits of a stable river, with less flooding and increased angler access, more than balanced the likely effects of a reduction in flow. The Tongariro would still be the largest tributary entering Lake Taupō, and would continue to draw many trout.[9] Ian Logan, also a fishing tackle dealer, was in favour of the project although some of his customers opposed it.[10]

Some local Māori also welcomed the power scheme, because it offered the prospect of secure employment for the region. John Asher, secretary of the Tūwharetoa Trust Board (and the Taupō county chairman), said the board was elated at the potential for local Māori, and added:

> Speaking on behalf of the owners, I can say that we are prepared to give every possible co-operation to the authorities to achieve the project's aims.[11]

The officials in the Wildlife Branch, outwardly at least, were largely in favour of the power scheme. In September 1963 Pat Burstall, Assistant Conservator of Wildlife, gave public assurances that the Wildlife Branch and the Marine Department had worked with power engineers from the start to make sure that no damage was done to the upper Tongariro fishery. He considered the hydro scheme had the potential to give better fishing because it would mean fewer floods, a smoother, more continuous flow and improved angler access, particularly to the lower reaches near the delta.[12] The engineers had guaranteed a minimum flow at the river mouth of 1,000 cusecs (28.3 cumecs) — the least amount necessary for trout spawning and good fishing. The natural mean flow of the Tongariro at Tūrangi at that time was 1,860 cusecs (52.6 cumecs).[13]

> If we were not completely sure no damage would be done, the [Wildlife] department would strongly protest about this scheme, he said.[14]

Hands off the Tongariro!

The green light for the Tongariro power development scheme prompted a stiff rebuke from a prominent New Zealander three months later. Peter McIntyre, a highly respected landscape painter, published an article calling on all citizens to oppose the destruction of the Whanganui, Whakapapa and Tongariro rivers. Although a Wellington resident, he was personally affected by the scheme because he had a holiday cottage at the tiny village of Kākahi, near Taumarunui. The cottage was superbly sited, overlooking a pool on the Whakapapa from which he had taken many a fine trout.

McIntyre gained his formal training at London's Slade School of Art. Students there came from all backgrounds; McIntyre later said it was the only true democracy he found on Earth. By the outbreak of World War II, he had spent nine years overseas, eking out a living as a bohemian artist. In 1939 he joined a New Zealand anti-tank unit with the lowly rank of gunner, before General Freyberg, commander of the New Zealand Division, appointed him New Zealand's official war artist. The appointment gave McIntyre a roving commission, allowing him to roam the entire theatre of war, painting men in combat or performing the hundreds of mundane tasks required for the conflict. At the war's end he returned to New Zealand, and began to record its people and places in his landscapes.

However, in his travels overseas, McIntyre had seen the environmental damage that industrial development had caused, and well knew what New Zealanders stood to lose from destructive public works. In arguing vehemently against the Tongariro proposals, he sought to shake the New Zealand public out of their complacency.

McIntyre was never one to pull his punches. He savaged the Ministry of Works plans for the Whanganui watershed, and started a war of words with Tom Shand, the Minister of Electricity. Surely, said McIntyre, the stage had been reached where the scenic assets being destroyed far outweighed in value the advantages of the energy scheme? New Zealand was in the grip of a 'mindless rush towards an industrialized nightmare State . . . the Philistine with his bulldozer is on the rampage'.[15]

McIntyre's protest was published in national newspapers in late June 1964. On the same day engineers from the Ministry of Works announced that they feared a 'power gap' would emerge in 1970 if the Tongariro power scheme was not speeded up. The projected completion dates for the first three stages of the scheme would be brought forward by two years. In making this announcement, the engineers seemed to be speaking as if government approval for the Tongariro power scheme was a mere formality.[16]

A few days later, the Minister of Works, Percy Allen, outlined what the future might hold for the tiny village of Tūrangi. The local iwi had agreed to sell to the government the land on which the new town would be built. This meant that a permanent centre could be constructed with many of the usual commercial undertakings, government offices and social facilities. There were even plans to convert the single men's hostel into a motel once the hydro scheme was completed. The town was to be

Peter McIntyre displaying one of his artworks.
Ref: 1/2-177202-F. Alexander Turnbull Library, Wellington, New Zealand. /records/22804621

located to the west of State Highway 1 and south of the Tongariro River bridge. This placed it well away from the original Tūrangi village site in Taupahi Road, hard by the river.

The government was still keeping to the story that construction work would not start until further consultation had been held with all parties likely to be affected, and decision-makers were better informed on how the interests of those in the region could be protected. That promise must have had a hollow ring to many. No one knew for certain how the trout fishery in the Tongariro would be affected, and the confidence of the officials in the eventual outcome might well prove to be misplaced.

Stung by the pace of these developments, Peter McIntyre formed the 'Hands off the Tongariro and Surroundings Movement' at the end of June 1964. In his capacity as the lobby group's newly elected president, he argued that New Zealand's natural scenic assets, its native forests and clear, free-flowing rivers were worth preserving for their own sake. The use of atomic power for electricity generation in the near future would make hydro-electric power schemes obsolete.

Destructive power schemes aimed at the Tongariro and Whanganui watersheds, when coupled with public apathy, meant that 'The Philistine is doubly armed today'.[17]

McIntyre hoped to convince the country's 130,000 sportsmen and women (his wife also fished for trout) to oppose the scheme, and planned public meetings to encourage them to join in the crusade. The movement was not overtly political, but more resentful of the government's apparent secrecy and resistance to public protest. Remarkably, there was no form of legal action in the courts. Administrative law — which allows the ordinary citizen to challenge government officials by questioning the way in which decisions are made — had not yet developed into the potent weapon it is today.

The movement's comments on 'rampaging Philistines' provoked a response from the Minister of Electricity, the Hon. Tom Shand. In questioning the accuracy of McIntyre's accusations, which he thought were in extremely bad taste, Shand tried to paint a picture of a hydro-electric scheme that had incorporated a strong conservation focus from the start.

Shand candidly admitted that water flows in the Whakapapa River, where McIntyre's cottage was sited, would be seriously depleted. However, major public works could not proceed without harming the interests of individuals.[18] Shand's arguments became personal when he suggested that, as a politician, he was not qualified to judge the worth of McIntyre's painting, so by implication McIntyre was unqualified to comment on complex public works. He also hinted that McIntyre was motivated by a desire for compensation for the loss in value of his holiday home.

In early July 1964 McIntyre gained an important ally. The aptly named John T. Salmon, associate professor of zoology at Victoria University of Wellington, agreed that McIntyre's objections were properly made because a public works scheme the size of the Tongariro should concern all responsible New Zealanders.[19] Salmon was an entomologist who had long been concerned about the plunder of New Zealand's wild and scenic areas, including rivers, by government departments. A book he had written in 1960 charged the Ministry of Works and the State Hydro-Electric Department (as it was then called) with being the two greatest menaces to New Zealand's wilderness scenery and the principles of conservation.[20]

With regard to the Tongariro power scheme, Salmon condemned

The Hon. Tom Shand, Minister of Electricity and opponent of Peter McIntyre.
http://livingheritage.lincoln.ac.nz/nodes/view/1120#idx7834. Obituary.

the government's undue haste, accusing it of following 'the usual steam-rolling tactics which the public has come to recognise as the normal pattern by which such schemes are hastily formulated and rushed through official approval'.[21] Tom Shand did his best to defend the consultation process, pointing out that the government stood by its assurances that no construction would commence until all reasonable measures had been taken to hear and consider objections.

Few Tongariro anglers could have believed this statement, coming as it did straight after the government's decision to bring forward the starting date for the scheme. Shand announced on 9 July 1964 that, because of an unpredicted rise in the nation's electricity demand that winter, a decision was required earlier than expected in order to avert an emergency in the government's power construction programme. An interdepartmental planning committee had been making forecasts of electricity consumption and peak demand for five-yearly periods,[22] but Electricity Department officials did not explain why they had not foreseen the possibility of a power shortage. The crisis was blamed on the spread of television (then a new form of popular entertainment) and New Zealanders' preference for electric bar heaters over kerosene-fuelled ones.[23]

There was a strong suspicion that the minister was papering over the cracks. TALTAC then came out with a statement formally opposing the power scheme, while Peter McIntyre argued that the days of hydro-electricity were over. Those anglers not against the project, such as TALTAC's Geoff Sanderson, McIntyre considered 'the most deluded fishermen in New Zealand'.[24] He called for an inquiry by an independent body into the merits of the current power scheme as compared with nuclear power. Ironically, the government Power Planning Committee had issued a report in early August 1964 calling for the use of nuclear power generation in New Zealand by 1977.[25]

The Nature Conservation Council

In seeking to give the Tongariro hydro-electricity scheme a veneer of respectability, the government drew on the opinion of the Nature Conservation Council. The council was a body created in 1962 by Act of Parliament to advise the government on conservation matters. Some of its members, drawn from the scientific community, were of the highest repute, and included as chairman Dr Robert Falla, Director of the Dominion Museum, and Peter McIntyre's supporter, Professor John Salmon.

The council had limited powers as an environmental watchdog, because it reported to the Minister of Lands in a purely advisory capacity. It was not a tribunal, it did not hear sworn evidence from witnesses, nor did it have the power to make decisions affecting anyone's rights or duties. Nonetheless, Dr Falla stated that: 'We have a semi-judicial function and will not pronounce judgment until we have all the facts we can get.'[26]

When the council held a hearing in September 1964 on the merits of the Tongariro power scheme, its members were divided on whether the scheme should proceed. A minority were totally opposed, while the others thought it could go ahead subject to modifications designed to protect the river. John Salmon had already written a newspaper article stressing the high recreational value of the Tongariro fishery and underlining the need for in-depth research on the effects the scheme would have. He was extremely concerned at the government's inadequate research:

"Oh, don't wait for ME!"

The Nature Conservation Council was confronted with a fait accompli. Cartoon by Gordon Minhinnick.
The New Zealand Herald, 20 August 1964

> . . . though assurances . . . from official quarters pour forth like water from so many taps, there exists no solid evidence to support and substantiate the claims that the amenities can be preserved.[27]

Frank Newcombe, Controller of Wildlife, represented the Department of Internal Affairs at the hearing. He told the council of the assurance given by the Electricity Department in 1955 that the Tongariro fishery would be safeguarded. More importantly, Newcombe stated that the undertaking had been carried out. He followed up by detailing the written assurances from the Minister of Electricity as to the maintenance of sufficient flows, artificial freshes, exclusion of the tainted Whangaehu river water from the scheme, and provision for future changes to the scheme as it progressed. Overall, he considered that trout would continue to run up the river and the fishing would be maintained.

The council voted by majority to pass a motion recording that, having weighed the need for extra electric power by 1972 against conservation factors, it 'offers no objection to the Government proceeding with the scheme, provided that every effort is made to preserve the recreational value of Tongariro River and, in particular, the proposal to build a new town at Tūrangi be reviewed with a view to siting it elsewhere'.[28]

Professor Salmon voted against the motion, while the chairman, Dr Falla, abstained. The result was then released to the press — no doubt to the relief of government ministers, but to the disgust of Peter McIntyre, who felt the Tongariro power scheme would ensure those of his generation would be viewed by history as vandals.[29]

The Tongariro power scheme begins

The Nature Conservation Council's decision meant the balance of power swung over to the engineers. The well-being of the Tongariro trout fishery was now in the hands of those who were to construct the power scheme, even though no coherent plan existed to carry out the undertakings that the government ministers had so confidently given.

In late September 1964 the government authorised stage one (the western diversion of Whanganui River tributaries), stage two (the Tokaanu power station), and stage three (diversion of Moawhango River headwaters and other streams into the upper Tongariro River). The estimated cost was £46 million. Work also began immediately on the new town of Tūrangi, which was to be located at the site originally planned. The government did not accept the Nature Conservation Council's recommendation for a different site, reasoning that urban development would have begun in the area even without the power scheme. The plans for Tūrangi envisaged the construction of more than 1,200 houses, along with permanent industry, a shopping mall, churches, primary schools and a hotel — all in accordance with the best modern town-planning principles.

The power scheme's destructive effects were soon felt on the upper Tongariro, and with it the lack of co-ordination between the local authorities involved. In October 1964 a contractor licensed by the Lands and Survey Department bulldozed an acre of bush at the Cliff Pool to install a crusher, cutting a swathe 1 chain wide down to the Dreadnought Pool bluff and slicing through the stopbank. The damage to one of the most popular pools on the upper river aroused the ire of Vice-Admiral Harold Hickling, by now a household name as the author of the most well-known book about the world's most famous trout river. Hickling slammed those responsible, condemning the work as largely unnecessary. The people of New Zealand, he thundered, were being robbed of a priceless heritage behind their backs, and the New Zealand government was aiding and abetting the process.

This furore, reported in detail in daily newspapers, forced an immediate government response. In a joint statement, the Minister of Electricity and the Minister of Works reassured the public that the commitment to protect the fishing qualities of the Tongariro River would be honoured both in spirit and letter.[30]

But there could be no turning back. In 1965, contracts were let for the tunnels on the Western Diversion and the Moawhango dam. An Italian concern, Codelfa-Cogefar, contracted for this work, which meant that a large part of the workforce would be from that country, subject to arrangements with the trade unions representing New Zealand labour interests.

Whakapapa River flow

Peter McIntyre's beloved Whakapapa River virtually ceased to exist after stage one of the Tongariro power scheme began, affecting rivers flowing from the western slopes of Mt Ruapehu. With most of its flow gone, the once beautiful runs and pools of the Whakapapa turned into a dried-up riverbed which soon became covered in willows, brambles and weeds. The desecration of a wild and scenic river was a curious result, because, under the Shand agreement governing the hydro-electric scheme, flows in the Whanganui and Whakapapa rivers were not to be reduced to levels at which the safety of fish would be endangered.

It was not until flows were partly restored in 1983 for recreational purposes that Kākahi residents, led by long-time campaigner Keith Chapple, felt things had returned to something approaching normal. After much litigation, a continuous minimum flow of 3 cumecs was fixed for the Whakapapa River below the Tongariro power scheme intake.[31] The decision was important for its inclusion of Māori values with regard to water resources.

Fishery protection measures

For Tongariro River anglers the future of their fishery depended on whether Pat Burstall and his colleagues could limit the damaging effects of the power scheme to the greatest extent possible.

Burstall then threw himself into the task of supervising the many fisheries aspects arising from the power development. If the Tongariro fishery were to be protected, then he was the right man for the job, for he had fished Lake Taupō as a boy. Arriving in Tūrangi after World War II, he had worked at the trout hatchery, intending it to be a temporary post before attending university, but the fisheries work attracted him, and he took up a permanent position as a fisheries technician. He was to spend eight and a half years at Tūrangi, accumulating experience in fisheries management and gaining an extensive knowledge of the Tongariro fishery. Although he had no formal scientific qualifications, as we saw in an earlier chapter he co-operated with Marine Department scientists in collecting and analysing data about the Tongariro River, and in doing so developed his own theories about the way the fishery worked.

Burstall now realised how vital it was for the various government departments involved in the power scheme to co-ordinate their efforts closely and consult with each other on a regular basis. He had seen the draft report prepared by Cedric Woods and disagreed with various conclusions, including the downstream migration of trout fry and the effects of a reduced river flow. Within two weeks of J. T. Gilkison's inconclusive meeting with angling groups in Wellington, he provided the Controller of Wildlife with his main concerns and suggestions.

To protect spawning trout in the Tokaanu Stream, Burstall proposed that the upper reaches should be fenced off. At the peak of the spawning season up to 2,000 trout could be present in the shallow waters, making them vulnerable to poachers. Burstall had no illusions about the likely fate of the fish were they to be left unprotected after a town of 6,000 people was erected in the vicinity. The Tokaanu Stream was also an important point for the collection of ova, although the Tongariro River was the main source.

Metal extraction from the Tongariro and other rivers posed another serious problem. The power scheme required an estimated 1.5 million cubic yards of aggregates. Much of this would have to come from riverbeds, causing continual discolouration of the water and difficulties for anglers. Burstall found that a metal extraction permit had been granted for an area near the junction of the Tongariro and Poutu rivers — a serious threat to the fishery in that locality. The Wildlife Branch had not been consulted on the matter, although on other occasions it had been. The omission arose from the sheer number of interested parties — Taupō County, the Lands Department, the Department of Justice, the Ministry of Works and the Waikato Valley Authority — all operating in an unco-ordinated fashion.

The vexed question of whether the Tokaanu tailrace should have an outfall into the Tongariro River or Lake Taupō was finally settled in early December 1964. When Pat Burstall met representatives of Sir Alexander Gibb and Partners and the Ministry of Works, it was explained that the addition of the tailrace water to the Tongariro's base flow would cause flooding in the lower river. The outfall would have to go in at Waihi Bay.

Pat Burstall obtained an agreement from the other departments that dredging could be necessary at the delta to maintain the quality of angling. It was another example of how he tried to anticipate every eventuality in order to protect anglers' interests.

Despite the large scale of the engineers' proposals, Burstall retained an optimistic view on the overall effects of the scheme. Key considerations were the short-term and long-term effects on angling and the fishery, and the effects of the power scheme on conservation interests in the region. He felt that if all government departments and others involved worked closely together, appropriate compromises could be reached and some good results could be achieved.

He also displayed commonsense when design aspects of fishery problems arose. It was necessary to set up electrical structures in the Tongariro River to count trout spawning numbers. In 1965 Burstall was sent to the United States to discuss this and other fisheries matters with wildlife officials there. On his return, the Electricity Department advised that it would be an expensive project involving diversion of the river. However, he was able install an experimental barrier near the hatchery using battleship chain and rubber mat.[32]

In the end, no amount of scientific research could accurately forecast the precise effects of the government's hydro-electric project on the Tongariro River. For both anglers and trout, only time would tell.

CHAPTER 19:

The effects of the Tongariro hydro-electricity scheme

'Tongariro will never be the same', Nature Council regrets.
— ***THE EVENING POST*, 24 SEPTEMBER 1964**

EVEN BEFORE CONSTRUCTION BEGAN on the Tongariro power development works, it was clear that the trout fishery would be affected to some degree and that those changes — mostly in the form of a reduced river flow — would be permanent. Wildlife Branch staff at the Department of Internal Affairs therefore had to ensure that unnecessary damage and disruption did not occur in the short term while preserving a viable trout fishery for the future.

To attain that objective, they had to understand how the Lake Taupō trout fishery functioned and the role of the Tongariro River within that fishery. Fortunately, surveys of anglers and their catches had been carried out since the 1950–51 season on a rudimentary basis, usually by rangers checking and recording during the fishing season. The experimental winter fishing seasons undertaken in 1955 and 1956 also required extensive inquiries into the trout harvest for those periods. With this information to hand, Pat Burstall and his colleagues could start to deal with the engineers' requirements in the initial phase of the power scheme's construction.

After recreational groups were consulted in 1964, Sir Alexander Gibb and Partners suggested that a comprehensive statement be prepared on the fisheries aspects of the power scheme. Pat Burstall wrote a detailed report in April 1965 covering a wide range of matters:[1] metal extraction, water turbidity, liaison with the Waikato Valley Authority, maintenance of minimum river flows, protection of trout spawning in the Tokaanu Stream, installation of screens to prevent mixing of fish stocks, the Tokaanu tailrace pollution problems, and better access for anglers to the Tongariro River.

The document was intended to be used as the basis for letting tenders, bringing fisheries protection measures to the fore in planning.

Cedric Woods has doubts

Despite the existing fisheries data, more research on the Tongariro was needed if the effects of reduced river flows, artificial fluctuations in river levels and other changes were to be offset. In early 1965 Cedric Woods, author of the technical report on the fisheries aspects of the power scheme (outlined in Chapter 18), confirmed that view when he noted that efficient fishery management required research, and that research was not being done.[2] His call was answered some months later when the Marine Department approved the erection of a laboratory at the Tūrangi trout hatchery. Woods also made the surprising comment that the government, by saying it was unlikely the fishing in the Tongariro would be harmed, had drawn too much from his conclusions. His report dealt only with problems that had to be answered before the power scheme plans were approved, and did not warrant any statement about what would happen to the fishing: it was only a statement on what could be done.

Woods stressed the report contained two important reservations: construction must not deplete trout stocks, and the fishery must be managed efficiently. 'The optimistic Government statement cannot stand unless full weight is given to these two most important qualifications.'[3]

It is probably no coincidence that Woods made these disclosures after he had resigned from his post at the Marine Department in early 1964. An expert's report in the hands of a politician is often treated as a smorgasbord of data, liable to selective interpretation. Woods must have been naïve in thinking the politicians of the day would not see his report as a full endorsement of the Tongariro power scheme. He had qualified his conclusions by noting that the report 'deals with the major fisheries' problems which require to be answered before the Power Scheme plans are finalised, and omits mention of other problems which can better be answered during or after completion of structural works'.[4] But whatever qualifications and contingencies he added were omitted or downplayed in ministerial statements that the fishing in the Tongariro River would not be affected.

However, the Minister of Works, Tom Shand, had given an important written assurance that fisheries protection was a factor in the power scheme operating procedures, which would be 'modified where necessary in the light of experience'.[5] The Electricity Corporation and the Ministry of Works made genuine efforts during the construction period to obey this injunction.

Looked at in retrospect, the forecasts by Woods seem to be very much a shot in the dark. The assumption that artificial freshes would induce trout to run up the Tongariro River in the winter months at times which best suited anglers was never properly tested. The Tongariro power scheme took some 20 years (1964–1984) before it was fully operational. During that period Wildlife Branch fisheries officers had to deal with problems as they arose. Only at the expiry of the construction period could the effects of the scheme on the fishery be analysed thoroughly and the appropriate management methods worked out.

The Tongariro power scheme was developed in several stages. It began with opening of tenders in April 1964 and the engagement of a large workforce, including specialist tunnellers from Italy. Slowly, Tūrangi was transformed from a quiet fishing hamlet into a small town. The first worker's house was occupied in mid-1965, and a tavern, a shopping centre and other facilities were to follow, but during the years in between Tūrangi was a construction camp with a distinctly frontier edge to it.

Residents of Tūrangi and Taupō had differing views on how the hydro scheme would affect the mighty Tongariro River. Some with business interests welcomed it, viewing the matter as a short-term disruption the effects of which would subside with time. Others feared that the Tongariro's wild beauty would be lost forever.

Project work on the hydro-electric scheme began in the late 1960s.

The damming of the Tongariro River took place in 1966, and the Poutu Diversion began operating in 1973, siphoning water down a canal to the project's Western Diversion. At that point the old Tongariro hands noticed the wild and beautiful river they had known was no more.

The most obvious indication of change was a reduced flow rate and, consequently, smaller fishing pools. The table below shows the Tongariro river flow rates pre- and post-1973.[6]

Location	Pre-1973 mean flow rate (cumecs)	Post-1973 mean flow rate (cumecs)
Puketarata Stream (upper river)	38.3	22.8
Tūrangi (lower river)	53.6	31.9

Beheading the upper Tongariro River also disrupted the seasonal flow variations which normally occurred with the change of seasons. The usual pattern was for April to have the lowest flows (25 cumecs at Puketarata) while the spring snow melt gave the highest flows in September (40 cumecs). But because the Poutu intake diverted water from the Tongariro, the increased spring flow had no effect downstream. As one researcher noted: 'seasonal flow patterns persist in the upper Tongariro and in the Poutu canal, but the average compensation flow in the lower Tongariro is constant throughout the year.'[7] Thus, supplementing the river flow to meet an agreed minimum for angling purposes was entirely artificial.

It also became apparent that the river's behaviour after a flood had changed. Previously it had taken almost a week for normal flow rates to resume, but the diversion at Poutu shortened that time to two days. Nature's processes were being turned upside down.

In July 1971, as turbid water flowed down the Tongariro River, local anglers responded angrily and accused the project works of causing the problem. Two senior engineers made an aerial survey and concluded that siltation had arisen from slips high in the catchment, but the anglers were not convinced.[8]

More than a year later the same problem surfaced again. This time, water from the newly excavated Tokaanu tunnel was seeping through banks of spoil left from tunnelling work and entering the Tokaanu Stream, bypassing silt traps intended to catch it. The stream's flow was also increased by 50 per cent as a result. To their credit, the engineers responsible took prompt action to pump the discoloured water away from the stream, but it was obvious that they and the contactors they had engaged had much to learn.[9]

Metal extraction benefits

Not all construction works had adverse effects. The extraction of metal from the bed of the Tongariro River helped improve fishing in some instances. In December 1964 the Ministry of Works decided to remove thousands of cubic yards of metal from the river below the main highway bridge. That quarrying took place at the sites of the old Swirl and Hut pools. Once work was complete the river was diverted back through the excavated beds, stabilising the river on the course it had taken before the 1958 flood. Not only was fishing improved, but by taking shingle from the river costs were also reduced.[10]

The Rangipo dam — silt and flow issues

The Rangipo dam, stage 4 of the Tongariro power scheme, is located on the upper river several kilometres above the Poutu intake. It is a run-of-the-river operation that takes water from the Tongariro River, the Waihōhonu Stream (a tributary) and the Moawhango diversion to an underground powerhouse which feeds the water back to the Tongariro via a tailrace tunnel.[11]

Rangipo was the first part of the Tongariro power development to be scrutinised in an environmental impact report, issued in July 1973. Official approval was given in November of the same year. It proved to be a difficult and expensive undertaking — something the government was well aware of — but clearly there were political considerations behind the decision. A skilled workforce had been formed during the decade in which other hydro-electric power stations had been built on the Waikato River, and the government wanted to keep this manpower resource fully occupied. Time was also running out, because the Power Planning Committee forecast increased electricity demand which would need to be met by 1970.[12] In August 1973, Acting Prime Minister Hugh Watt showed enthusiasm for the Rangipo scheme in an address to a conference of the New Zealand Contractors' Federation when he indicated finance was already arranged. The engineers were also ready to start work.[13]

From a fishery viewpoint Rangipo became controversial once work began in 1975. Because the scheme is beneath the ground and is operated remotely from Tokaanu, much tunnelling was required to construct surge chambers, access tunnels and drop shafts. Anglers noticed the dramatic effects as suspended silt from tunnelling activities made the Tongariro River's clear blue waters a murky off-white shade, unattractive to both anglers and trout. The discolouration was a constant feature for the 1977–1979 seasons, making the river unfishable six days out of seven for all but the most persistent anglers. (Sundays offered a respite because project work seemed to cease.)[14]

The murky flow was especially frustrating to the guides who worked the Tongariro River. Foreign clients who came to Tūrangi expecting to find the sparkling waterway shown in the tourist industry's advertising received a shock when they first encountered a river looking more like a sluice channel from a gold mine.

The silt problem soon hit the media. 'Grey's Tongariro now an Eldorado tarnished by silt' admonished local newspaper the *Taupo Times* in July 1978. At about the same time, O. S. Hintz, President of the Federation of Lake Taupo Angling, Shooting and Boating Clubs, wrote to the Minister of Works and Development, the Hon. William Young, to express anglers' concerns. The Rangipo powerhouse excavations caused sediment discharges from three sources, one of which was the tailrace tunnel work which sent its wastewater into the Tongariro River below the Poutu intake.

The project engineers had installed settling ponds which trapped some sediment particles, but the finer silt and clay was being carried through the system and into the river. The minister argued that even without the Rangipo scheme, the floods which occurred in the Tongariro up to five times each year could disturb fine sand and sediment and discolour the water, as well as moving sediment into pools lower down. The blame could therefore conveniently rest on allegedly natural factors.

But the real reason for the government's refusal to take wider measures (Hintz had argued for a chemical flocculation system) was simply the cost. The minister instead offered additional trout-rearing capacity at the Tongariro hatchery — an olive branch which the Department of Internal Affairs' Wildlife Branch accepted.

The reduced river flow had a definite influence on the catch rate of anglers. In the early 1970s the standard fishing technique was to cast a large wet fly lure downstream and retrieve it upstream. That method was less successful in low water conditions. Some upstream fishing was possible with a dry fly, but it was restricted mostly to the summer. The practice of upstream nymphing — later to become widespread — was yet to arrive. There was also evidence that fish tended to spawn in the main Tongariro watercourse instead of one of the tributaries.

The Tongariro power scheme was initially planned and completed by the government department responsible for all public construction projects, the Ministry of Works and Development. The New Zealand Electricity Department was responsible for operating the completed scheme. Later, following the wave of privatisation brought about by the Labour administration, the power scheme passed into the hands of state enterprise Genesis Power Ltd. When in 2000 Genesis Power applied for resource consents to operate the scheme, the Department of Conservation questioned the adequacy of the existing prescribed flow rate (16 cumecs constant flow below the Poutu intake on the upper river) and proposed increased flows more in line with those that had occurred in natural conditions. The resource consent application led to an agreed solution to meet Department of Conservation concerns, discussed below.

Flow management survey

In the early 1980s there were indications that the scheme as originally planned was not satisfying the requirement to protect the trout fishery. This failure was partly because the flow management regime was being implemented while the number of anglers fishing the Tongariro increased and the annual trout harvest grew larger. In 1983 the New Zealand Wildlife Service, with funding from the then Ministry of Energy, began a scientific study of the flow management regime. One key question was what factors influence the number of adult trout running up the Tongariro river. If that question was still unanswered, then, by implication, Cedric Woods's theory of using artificial freshes to prompt trout migration into the Tongariro River was not working.

The health of the Tongariro River became a media topic. 'Who's Killing the Tongariro?', a no-holds-barred series of articles by Trevor Cameron appeared in the *Turangi Chronicle* in July 1987.[15] A year later a national daily newspaper covered silt build-up in the river's pools in a brief article, 'Mighty Tongariro is Choking to Death'.[16] A Tūrangi rafting guide argued that the failure to provide sufficient flows to scour the riverbed was slowly ruining the Tongariro, with the prospect of no fishing at all in another 50 years. 'Is this the price we pay for progress?' he asked.[17]

Fisheries research methods had advanced enormously since the pre-war days of trapping migrating trout and recording elementary data from anglers' catches. Scientists carried out the six-year study by electro-fishing parts of the river to establish the numbers of juvenile trout present in different locations. (An electric current attracts the fish to an anode, which temporarily stuns them, allowing the counting and evaluation research to be conducted without harming the fish.) And because a trout carries its life history on its back, they could also analyse trout growth rates by studying their fish scales in much the same way as a botanist studies growth rings evident in the wood of a tree. The research indicated that many juvenile rainbow trout spent their young lives in the middle section of the river above Tūrangi or in the Whitikau Stream, one of the Tongariro tributaries, until floods forced them to the main river and thence to Lake Taupō.

But while many juvenile trout migrated or were swept downstream to Taupō's wide waters, surveys of trout numbers in the lake showed that only a small percentage survived. Therefore, adequate nursery habitat for trout fry and fingerlings was of vital importance. Any revisions to the power scheme flow regime would need to take account of that factor.

The 1989 study recommended that the minimum flow requirement fixed for the lower river at Tūrangi should be abandoned and replaced by a system of monthly minimum flow rates at a point upstream, below the Poutu intake. The lowest rates would be in summer (13 cumecs) and the highest would apply in winter and spring (16, 17 and 18 cumecs). The result would be fewer minor surges in the river level and a flow rate more in line with the Tongariro's natural flow. There would be more habitat suited to juvenile trout, with reduced incidence of sand accumulation on the riverbed — an important factor in trout food production. There was also an element of meeting the market; as the study noted, monthly adjustments to the river's compensation flow 'will give anglers higher flows during the angling season, which is what anglers seem to think they want'.[18]

John Gibbs, Taupō Area Fishery Manager, concluded that reduced flows in the lower Tongariro reaches meant less rearing habitat for juvenile trout. That in turn made for fewer sub-adult trout and a consequently smaller pool of fish contributing to the entire Lake Taupō fishery. A smaller trout population in the lake saw fewer mature fish entering the Tongariro River on their annual spawning migration, which made the fly fisher's bag smaller.[19]

Lower flows also had adverse effects on the general angling experience. No more could anglers feel the thrill of fishing a large, demanding and sometimes dangerous river, where both current and trout were strong. Gibbs noted:

> Instead of the challenge of a large, boisterous, unique river demanding a special approach and techniques, anglers are now faced with a smaller, tamer river, much more like those they can fish elsewhere in the country, or indeed elsewhere in the world.[20]

For those fly fishers who had known the river before its waters were diverted, the smaller pools and wider expanses of exposed riverbed produced an acute sense of loss. An informal survey which the Department of Conservation conducted in the late twentieth century supported that sentiment. Whole-season anglers were asked if they were happy with the state of the fishing on the Tongariro. Of the 44 anglers who responded who said they had fished the river before the hydro scheme began, most (84 per cent) felt that the power generation scheme had adversely affected the fishery in various ways.[21]

Despite efforts made to mitigate the effects of the hydro scheme, the quality of the fishing declined. The most obvious indication of this trend came in the form of smaller catches for anglers. Statistics compiled by Department of Conservation fisheries officers in 2000 confirmed the anglers' fears: before the Poutu intake diverted river water in 1973, the mean catch rate of fish per hour of angling effort was 0.42; from 1973 onwards the corresponding rate was 0.25.[22] That difference alone did not conclusively mean that the Tongariro power scheme had damaged the fishery. But when the officers used the same survey methodology for two other important Taupō rivers, the Waitahanui and the Tauranga–Taupō, they found no marked difference in anglers' mean catch rates for either river in the periods before and after the Poutu intake began operations in 1973.[23]

Another fisheries officer, Glenn Maclean, observed that the Tongariro River played an important role in the maintenance of the

Lake Taupō fishery. He noted the Lake Taupō trout population, and therefore the total number of trout caught, was greater in the 1960s and 1970s than in the 1990s.[24] Because the Tongariro River was the major contributor of juvenile trout, any interference with recruitment in that tributary would have major consequences on the fishery overall and would affect winter and summer catch rates. There were fewer mature trout running the river in the post-1973 years because recruitment from the Tongariro's main stem — the part of the river affected by the hydro scheme — had declined.[25]

The accumulating evidence suggested that since the hydro scheme began the Tongariro River did not function in the same way as other Taupō rivers. One pointer to this conclusion was the degree of fluctuation in trout spawning numbers which occurred in other rivers, including the Tongariro's tributaries. Similar variations did not feature in the Tongariro's main stem.[26]

Maclean's reasoning gained support from the results for the 2000 fishing season, when the Taupō fishery had an exceptional year. Numbers of mature fish surveyed in selected parts of Taupō rivers were very high, with correspondingly higher angler catch rates. But for the Tongariro the mean catch rate was only marginally higher, when it should have been at similar levels to those experienced in the 1960s and 1970s.[27]

The Genesis Energy hearing

In 2000 Glenn Maclean and his colleagues got the chance to put their views publicly when the resource consents granted to Genesis Power Ltd (renamed Genesis Energy Ltd) for the operation of the Tongariro power development scheme came up for renewal. Many individuals and organisations, including the Department of Conservation, made submissions to the local authority commission appointed to decide the matter. The department argued the minimum river flow below the Poutu intake (the lower Tongariro River) should be increased from a flat 16 cumecs to a range of rates which more closely followed the natural seasonal flows in summer and winter.

Maclean and other officers gave evidence of a decline in anglers' catch rates since the power scheme began, asserting that the reduced flows had an adverse effect by diminishing the rearing habitat for juvenile trout. However, the commissioners could find no clear link between the power scheme and the decline in the quality of the fishery. They suggested that the answer lay in the number of Taupō anglers (the Tongariro being the most frequently fished river in New Zealand) and over-harvesting of the trout fishery. These factors, said the commissioners in a rather speculative tone, 'may have had an influence on catch rates'.[28]

The Department of Conservation filed an appeal against this decision, but later settled the issue without further court hearings. So the 16 cumec minimum flow rate at the Poutu intake remained, resulting in a 27 cumec rate lower down the river at Tūrangi. Settlement was also reached with Genesis Energy Ltd on flow rates for the upper river and the Poutu tributary, as well as sediment flushing plans for the Rangipo dam.

The main effects of the power scheme have been to reduce both the amount of water in the river and the variation in river flow. Under natural conditions the mean flow at Tūrangi was 53.6 cumecs; once the scheme was in operation that rate reduced to 31.9 cumecs.[29]

Despite the changes to its flow, the Tongariro could still give up large trout while the hydro-electricity scheme was in the course of construction. As an indication, the records of the Tongariro and Lake Taupo Anglers Club show eight rainbow trout of 10lb or more were caught between 1960 and 1989.

Genesis Energy's mitigation measures

As compensation for the loss of recreational activities afforded by a wild and scenic river, Genesis Energy Ltd, the operator of the power scheme, is required to undertake a number of actions that will go towards mitigating its adverse effects. One of these is the requirement to divert water into the Tongariro River for three weekends per year, in addition to any water needed to give flushing flows.[30] (Flushing flows are designed to prevent the unwanted accumulation of periphyton on the riverbed.) Increased flows also have great attraction for white-water rafters and canoeists.

The company also aims to ensure that extra flow in the Tongariro is sent outside the legal hours for fishing on the river, although this practice is occasionally breached. A sudden spike in water flow from 30 cumecs to 60 cumecs in 2006 caused consternation among river users.[31] On 20 September 2017 the flow at Tūrangi more than doubled to 56 cumecs, with obvious risks for anglers who had crossed the river to reach a favoured pool. Significant increases in flow, usually as the result of maintenance requirements, are therefore signalled to river users in advance.

The company has also worked with the Department of Conservation and the Tongariro National Trout Centre on a programme to foster breeding and propagation of the native blue duck (whio) across New Zealand. In September 2011 it signed a five-year $2.5 million business partnership with the department to aid whio restoration.[32]

On 15 July 2020 the Department of Conservation announced there were 748 pairs of whio in security sites across New Zealand, an increase from the 298 pairs found across the eight national security sites in 2011, the time when the whio recovery programme began. Whio present a challenge for conservationists because they rely on clear, free-flowing rivers for survival. Unlike other bird species, they cannot be shipped to an offshore island to breed undisturbed.

By agreement with the Department of Conservation, contributions have also been made annually towards trout fishery enhancement on the Tongariro River via the Tongariro Fisheries Enhancement Fund.

Fishery management strategies

The Tongariro power development scheme has undoubtedly altered the trout fishery, but one of the beneficial effects is the proliferation of scientific research that has resulted, much of it funded by the scheme's operator, Genesis Energy. Few trout fisheries anywhere else in the world have been the subject of such detailed studies for such a lengthy period. Knowledge obtained from this research is potentially relevant to rainbow trout fisheries outside the Taupō region.

Department of Conservation officials are constantly monitoring the state of the trout fishery at Taupō, and apply different strategies to provide the best possible angling experience. Experiments influencing the timing of the spawning runs of Tongariro trout form part of their work.

Restoring the early run of rainbows

The migratory habits of Lake Taupō's rainbow trout have changed significantly over time. In its initial stages, the Lake Taupō fishery was characterised by a spawning season that was shorter, and more concentrated, than the spawning runs we see today. Given the right conditions of weather and water, rivers almost devoid of trout would the next day be filled with fish, at least in their lower reaches. In the 1930s Waitahanui Lodge owner Fred Fletcher, who kept a close eye on his local water, noticed a rainbow run start after dark, and became one of the first anglers to take advantage.[33]

In the decades after World War II the spawning season began to lengthen, with more fish running in early winter (May and June). This change began to inconvenience anglers, whose lawful fishing season on the Tongariro River ran from 1 December to 31 May.[34]

The experimental winter season followed in 1955, after which the fisheries managers adopted the 12-month season now in force (1 July to 30 June). But the question of why most trout failed to run the rivers in autumn remained.

Tongariro trout surveys

Freshwater fishery scientists use fish trapping as a method of assessing the current state of the Taupō fishery and making forecasts of future trends. A fish trap was first set upon the Waipa Stream, a Tongariro tributary, in 1998. After four years, Department of Conservation officers gleaned useful data on the size, condition and numbers of fish running the river (both rainbows and browns).

The data showed that the 1998 run was exceptional, with the average weight for rainbows (5.29lb) being markedly heavier than the corresponding figure for 2001 (4.14lb). A possible reason for the increase was the effects of ash entering the Taupō fishery from the local volcanic eruptions occurring in 1995 and 1996.[35]

Most of the fish migrating up the Tongariro were three-year-old rainbows making their maiden spawning runs. A reasonable proportion — one-third of the run in 2001 — were fish spawning a second time. Statistics for brown trout revealed a wholly different picture for the 1998–2001 period. Second-time spawners were roughly equal between the species at around 20 per cent, but only 1 per cent of rainbows spawned a fourth time compared with 16 per cent of browns.

The tables opposite illustrate the numbers of rainbow and brown trout, and their average weight, passing through the Waipa stream trap in the four years 1998–2001.

Species and sex	1998	1999	2000	2001
Rainbow female	1,949	3,666	4,109	4,229
Rainbow male	1,151	2,451	2,707	2,261
Rainbow total	3,100	6,117	6,816	6,490
Brown female	312	287	413	381
Brown male	257	157	257	292
Brown total	569	444	670	673

Average weight (kg)	1998	1999	2000	2001
Rainbow male	2.29	1.68	1.85	1.77
Rainbow female	2.48	1.75	2.02	1.94
Rainbow total	2.40	1.72	1.95	1.88
Brown male	3.39	2.56	2.84	3.08
Brown female	3.17	2.59	2.75	2.89
Brown total	3.26	2.58	2.79	2.97

Source: *Target Taupo*, March 2002, Issue 39, pp 26–27

Two points arise from these survey results. First, they illustrate the large fluctuations which can occur in the Taupō trout population from season to season, with the number of rainbows running in 2001 more than twice the number passing through the trap in 1998. Secondly, while there were fewer rainbow trout in 1998 than in 2001, their average weight was greater, as noted above. The 1998 rainbow trout weights suggest that with a smaller trout population, each fish entering the lake and reaching maturity obtained a greater share of the available food supply, assuming that supply did not fluctuate markedly.

The early-run experiment

In the early twenty-first century, surveys carried out on the timing of the trout spawning run in the Lake Taupō tributaries showed the majority of fish were migrating in spring (September to November).

Survey data for the period 1987–2002 indicated that the main run on the Tongariro would usually begin after the first significant rains in mid-June.[36] This contrasted with the erroneous belief held by many anglers that the winter run began earlier, at Easter. In 2011, fisheries scientist Glenn Maclean noted that the rainbow spawning run had become later during the previous decade. The timing of the run had moved away from the April–July period.[37]

Spring spawning could cause problems for both fish and anglers, because it meant the trout returned to the lake later in the year. That meant they had less chance of feeding on smelt shoaling in the shallows during the summer months — a rich food supply, and one essential for trout to regain their former condition. Trout feeding near the shoreline were the easiest for anglers to catch, so anglers' bags contained a high proportion of poorly conditioned fish.[38]

The main spawning months for Tongariro River brown trout had remained the same (May to July). The earlier spawning of the browns meant their progeny faced no competition from rainbows when they hatched. Juvenile browns could be first to take over prime territory in the river and would be larger when any rainbow fry arrived later. Thanks to this head start, brown trout could realise their full spawning potential.[39]

Department of Conservation scientists considered whether to alter the fishery to make fish spawn earlier in the year — the so-called 'early run'. That meant stocking the wild fishery with trout that had early-run propensities. Artificial stocking is controversial among fisheries scientists because it raises a host of issues. Mixing wild and stocked fish could bring undesirable consequences no matter what the scientific modelling projected. Experience in other parts of the world had shown that stocking failed in fisheries where natural spawning took place. The fundamental goal had to be the preservation of the genetic integrity of indigenous Taupō trout.[40] Even if all the signs were favourable for a stocking programme, the results would take many years to emerge, which meant long-term monitoring.

Possible sources of rainbow trout with the early-run characteristics were fish from Lake Rotorua or — nearer to Taupō — Lake Otamangakau. Department of Conservation scientists experimented by raising trout from ova taken from trout resident in both lakes. They released some 5,000 young trout (10cm or more long), but with no tangible result — either because the fish were too few to have an impact on the trout population or, being hatchery-bred, they lacked the inherent survival characteristics of wild trout.

Later studies pointed to the size of the fish as being a significant factor in the timing of the annual runs.

Angler harvest — CPUE analysis

Despite technological advances, one of the most important tools by which fisheries scientists manage a fishery is to record the size and quality of the trout that the anglers catch. Statistical analysis of properly obtained data can reveal trends operating within the trout population which would otherwise remain obscure.

A key indicator in this context is the catch per unit effort (CPUE); that is, the number of mature trout that anglers catch for each hour they spend on the water.

A factor affecting the CPUE is the range of skills among anglers. A small minority will be in the expert category and will take a higher proportion of the annual catch. At the opposite extreme are the newcomers to the sport, who, if they have any success at all, can ascribe it to chance and take far fewer fish. Also relevant are the numbers of anglers participating each season (the Taupō fishery is open to all comers) and the estimated number of trout making the annual spawning run.

A question for scientists is whether the trout population can alter markedly without any significant change in the CPUE — a phenomenon known as 'hyperstability'.[41] In that situation the CPUE is not always a reliable indicator of the Tongariro River trout population at any particular time.

In 2015 Department of Conservation scientist Michel Dedual used archival angler catch records for the years 1948–2013 to investigate whether hyperstability might exist in the Tongariro River. Trapping data for a Tongariro tributary gave estimated trout numbers for the 15 years from 1998 to 2013.

The study concluded:

> From a management point of view the most significant result of this work is that between 1998 and 2013 fish abundance had very little effect on CPUE . . . on a yearly basis or on a monthly basis . . . ; even when the index of fish abundance was changing by a factor of 4. This

> strongly suggests that some hyperstability in CPUE occurs in the Tongariro River. The exact mechanisms driving hyperstability in wild trout fisheries are not well understood but we know now that CPUE can increase — not because there are more fish but because there are fewer unskilled anglers. This means that the pattern of participation in angling can lead to hyperstability.[42]

The analysis indicated that for the Tongariro trout fishery between 1985 and 2013:

- the level of catch inequality declined
- the average CPUE increased
- total licence sales declined
- the number of fish available to anglers declined.

The decline in catch inequality was surprising, because forecasts had projected that with fewer fish in the river, catch inequality would increase. Possible causes of this result were an improvement in angling methods and equipment, and changes in the type of angler fishing the river — economic downturns made more novices give up the sport, leaving a core of expert angling addicts:

> The increase in CPUE, together with the decline in participation, supports the hypothesis . . . that when unskilled (short-term licence holders) anglers are the first to abandon fisheries, a higher proportion of experienced (season licence holders) anglers will remain. Therefore, any change in the number or proportion of unsuccessful anglers will affect the level of inequality and the CPUE.
>
> This means that if we (as managers) aim to reduce catch inequality we have few effective actions available. Taupo fishery is an open-access fishing meaning that anybody (skilful or not) can legally fish; we cannot control which type of angler is leaving or joining the fishery.[43]

The study concluded that if CPUE was taken as a measure of angler satisfaction, things had generally improved in the period under review:

> The objectives when managing recreational fisheries are primarily to sustain or improve the quality of fishing so that anglers remain satisfied, and to sustain the fish population. Anglers can be satisfied

for several reasons, but maintaining or increasing CPUE generally maintains or increases satisfaction. Therefore, if we assume that CPUE is an important measure of the fishing quality, then we can conclude that average fishing quality has improved between 1985 and 2014. [44]

Radio-tracking surveys

From time to time Department of Conservation scientists have used electronics to solve some of the mysteries surrounding trout migration. There are some basic questions which are of interest to anglers. What is the peak for the annual spawning run? How long do trout remain in the river? Which parts of the river are used for spawning? What is the significance of the Tongariro tributaries?

In 1995 scientists sought answers by trapping trout near the Tongariro River mouth and implanting them with small radio-transmitting units. Researchers then walked along the riverbank with telemetry equipment. In this way the location of the fish could be found on a given day by following the pulses from the implanted tag.

The 1995 project gave up much valuable information by indicating the pools which trout especially favoured for spawning — useful knowledge for the angler. In the years that followed, the Tongariro River was affected by ash deposits from an eruption on Mt Ruapehu. Some pools disappeared after a large flood in 1998. These events prompted the scientists to carry out another tracking project in 2003 for an updated view of the state of the fishery.

Results in both 1995 and 2003 showed that a fresh in the river prompted an increase in trout average movement rates. These occurred when the river was rising and falling. The fish also used most of the Tongariro for spawning, from de Lautours Pool right up to the Poutu intake above the legal fishing limit. In 1995, fish were spread more or less evenly across the lower, middle and upper sections of the river. The 2003 study showed the trout had a distinct preference for spawning on the middle section (from the Breakaway Pool to the main highway bridge).

The trout also showed a preference for migrating between midday and midnight rather than the opposite. The presence of daylight also influenced them, with most fish movement taking place between 4pm and 8pm. Anglers can add these findings to their stock of knowledge in the pursuit of their elusive quarry.

Fisheries outlook

From the viewpoint of the angler, there seem to be two undeniable consequences of the Tongariro hydro-electricity project. One is the loss of a wild and scenic river regulated only by Nature. The other is a reduction in the size of the river.

For scientists, a smaller river means a loss of habitat that would otherwise be available to juvenile trout reared after hatching in the main river or a tributary. If juvenile trout (being territorial) are limited by their environment, there are fewer fish to migrate down to the lake to form the basis of the next year's age class.

Juvenile trout have also been found to exhibit poor growth rates in the Tongariro River. Research has indicated their diet has changed from one largely made up of mayfly and caddis fly to chironomid and oligochaete worms. The first two categories provided greater trout growth with less effort. Smaller trout reaching the lake at 18 months have less chance of survival in Lake Taupō.[45]

Against this, some freshwater scientists consider the changes wrought by the hydro engineers since the 1960s have improved the trout fishing. There is an ability to control flows and reduce, if not eliminate, the effects of floods and droughts. Flushing flows may reduce periphyton growth and alter insect life to the benefit of the trout.

CHAPTER 20:

The arrival of nymph fishing

This nymph fishing is a great way to take trout.
— **JOHN PARSONS, *PARSONS' GLORY, A BEDSIDE BOOK FOR ANGLERS***

ALTERATIONS TO THE TONGARIRO RIVER'S FLOW were not the only changes on the angling scene in the 1970s; there were changes in fly fishing methods as well.

Shooting-head fly lines

The first to arrive, courtesy of United States anglers, was the shooting-head system, in which a very heavy but relatively short length of fly line was attached to a light braided backing. Once the fly line was through the top rod ring, its weight could be made to propel the lighter line out across the water for an impressive distance, provided the timing was right. This method was an improved way of getting the lure down to the fish.

Fishing guide Tony Jensen first learnt about this method in 1969 when he took Joe Brooks, an American client, to the Island Pool. Brooks was well used to fishing for steelhead in his home state, and immediately saw that the vital factor for success on the Tongariro River was to get the fly down near the bottom, where the fish were.

A shooting head was the solution. Brooks had no stripping basket, but a cardboard box from a local grocery store served as an adequate substitute. After replacing his fly line with a high-density sinking head backed by a length of floating line it was back to the river and game on. This time the fly worked right on the riverbed, skimming the boulders, and a beautiful 6lb rainbow was the result.[1]

The nymph technique

By 1974 the wet fly anglers progressing downstream in many pools noticed there was angling traffic coming up-river towards them, casting busily and watching the water intently. And these upstream merchants weren't merely fishing; they were catching fish. Nymphing had arrived on the Tongariro. Anglers using traditional lure fishing methods simply made room for the nymph fisher coming up-river. What had been a one-way street became a dual carriageway.

Nymph fishing had been common practice in New Zealand trout waters outside the Taupō and Rotorua regions for the better part of a century. The earliest fly fishers, such as Captain G. D. Hamilton, were exponents of the small wet fly, which worked well on the introduced brown trout. Later, in the early twentieth century, the sunk fly method of Englishman G. E. M. Skues became well known to Antipodean anglers, who began with standard English nymph patterns, such as Hare's Ear and Dark Olive, and later developed their own after observing fly life in their local streams.[2]

These nymphs were not weighted, because the angler usually fished to trout feeding just beneath the surface. The nymph was the perfect solution to the problem of 'bulging' trout — those difficult fish which appeared to be rising to surface insects but were in fact taking rising nymphs sub-surface.

John Parsons, a Tongariro regular, recorded that nymph fishing began on the river in May 1974 when Tony Swainson, a Lancashire angler turned New Zealand resident, took a Tongariro rainbow on a nymph fished upstream. Swainson showed the method to two others from the Taupō fly fishing club, an Irishman and a Yorkshireman, so the earliest converts to the nymph technique were a small cabal from the United Kingdom.

Hooking Tongariro River trout posed a problem, because they tended to lie near the bottom of the river's deep pools where irregularities in the current caused by boulders allowed them resting space. To get flies down to the trout's level, very long leaders — up to 16 feet in some cases — and heavily weighted flies were required. The insertion of lead or other material to make a weighted fly had been prohibited in Taupō waters for decades, although it had been legal in the Rotorua lakes district only an hour's drive away since 1973. The hook used for the fly could be no larger than size 10.[3]

These restrictions led to some odd results from the angler's point of view. Writing in *New Zealand Outdoor* magazine in 1976, a correspondent called 'WJC' weighed standard Taupō flies of different sizes along with a size 10 Pheasant Tail nymph tied with a body of copper wire.[4] The results were:

- Orange Rabbit (No. 2 hook) — 520mg
- Fuzzy Wuzzy (No. 4 hook) — 350mg
- Taihape Tickler (No. 6 hook) — 300mg
- Pheasant Tail (No. 10 hook) — 200mg.

Under the regulations, the Pheasant Tail nymph, the lightest of all, was illegal, while the three standard flies weighed more but were legitimate. 'WJC' felt the regulations, which had been drafted years before nymph fishing became popular, should be revised to take a more realistic approach. If anglers trolling for trout on Lake Taupō could use leadlines and large spoons to get down to the desired depth, why should there be a restriction on weighted nymphs?

> The angler casting the traditional Taupo lure across and downstream has the advantage of a heavier fly and a high-density line — the angler fishing the unweighted nymph with a floating line might as well sit and twiddle his thumbs on a river like [the] Tongariro. The Tongariro River is so deep and the current so fast he would be lucky if his unweighted nymph sinks more than a couple of feet . . .[5]

Perhaps the protests had some effect, for under new regulations issued in 1978[6] the definition of an artificial fly for the Taupō district was amended, allowing weight to be added to flies tied on a size 10 hook no longer than 14mm and with a gape not exceeding 6mm. Apart from this concession, nothing by way of lead, glass or plastic could be added to the angler's cast, line or fly to aid with casting or help sink the fly line.[7] In 1983 the legal hook size changed to the larger size 8.

The restriction on the size of the hook limited the amount of lead wire that could be applied to a Tongariro nymph. Anglers got around this problem by attaching a pair of beads from a bath plug chain above the hook eye, giving even greater weight to the fly. These bug-eye rigs became popular on the Tongariro River by the 1980s and certainly got the flies down deep, but the method was crude by comparison with ordinary fly fishing. Many people attracted to Tūrangi wanted to catch a limit in the shortest time possible, and using a 'bomb' seemed the best way to do that. Some were newcomers to fly fishing, ignorant of the rules of angling etiquette, and there were arguments on the waterside when crowding occurred at a known 'hotspot'.

Disputes over fly fishing techniques

A group of experienced Tongariro River anglers felt that the sport of fly fishing had changed to a form of float fishing. In 1987 they prepared a paper calling for the regulations to be changed to restore 'traditional flyfishing methods and values to fly-only waters'.

The Taupō Ward of the Central North Island Wildlife Conservancy Council, the statutory body representing fishing and hunting interests in the region, asked the Department of Conservation to look into the matter. When rangers made a survey of Taupō anglers in 1988 it became clear that some people had concerns. The fisheries officers concluded that while the heavyweight nymphing method would not adversely affect the trout population, it could affect the quality of the fly-fishing experience many Taupō anglers valued. That required a line to be drawn between legitimate fly fishing and the invalid float fishing method some anglers resented.

The Department of Conservation explained:

> Obviously anglers' interpretations of flyfishing will vary and no one definition will satisfy all anglers. Managers feel the definition should strike a balance between the desires of 'traditional' flyfishers and the sentiment that it should be possible for the majority of anglers to have a realistic chance of catching a fish.
>
> We think a useful working definition of flyfishing at Taupo is: 'The practice of fishing for trout using traditional fly casting techniques with a flyrod, reel, flyline and leader to fish flies which are either imitations

> of natural food or incite an aggressive response. These flies may either be presented downstream or allowed to drift naturally without additional weight or buoyancy other than that provided by the line or by the fly.'
>
> Applying this definition, the use of size 8 hooks onto which as much lead as possible is wound solely to act as a sinker is not 'flyfishing'. Neither is the use of a float which is necessary to support the end of the flyline when using such weights.[8]

In 1988, after canvassing fishing clubs and undertaking riverside surveys, the department had obtained a range of possible solutions from the angling community. The responses included banning the use of all strike indicators and weighted nymphs, restricting nymph fishing to certain areas, limiting the size of the hook on which a weighted fly could be tied, and permitting wool or yarn indicators only. Some of these remedies were obviously unenforceable, and the policy ultimately adopted came down to two requirements:

- indicators to be either synthetic or natural yarn, and
- weighted flies to be size 10 or smaller.

Because the Tongariro River is so wide and turbulent, strike indicators placed at the end of the fly line became essential if anglers were to easily detect a take by a trout. Any fly rod bearing a brightly coloured tassel was a sure sign that its owner was a nymph fisher, and they became a common sight on the river. Indicators were often so large and unwieldy that fly casting in the ordinary way became a distant memory.

The Department of Conservation noted the trend in 1989:

> In the continual quest for a more successful technique which intrigues most anglers, flyfishers experimented further with strike indicators made of many different materials and tried all sorts of different rigs. The extreme has been the development of rigs involving nymphs so heavily weighted that the end of the flyline pulls under unless supported by several polystyrene balls and which can only be fished by roll casting repeatedly.[9]

The ban on strike indicators made from materials other than wool or yarn began on 1 July 1990. The rule change seemed to have the desired effect; 'float fishing' disappeared and there were no complaints from anglers, a point noted by the Department of Conservation when it reviewed the fishery regulations in 2003.[10]

In 2017 the definition of fly fishing was refined to mean 'to fish for trout with a fly rod, fly reel, fly line, leader, and artificial fly or flies'. Under related definitions the fly line must be at least 3 metres long and the angler's leader (including tippet) must not exceed 6 metres.[11]

Arguments over trout fishing techniques are nothing new. The history of the sport reveals that trout fishers have disputed from time to time the validity of a particular fly pattern or the way in which it can be fished. For example, in the late nineteenth century one of the great fly fishing debates saw F. M. Halford, the English dry fly purist, cross swords with G. E. M. Skues. Skues was a highly observant man whose careful study of the feeding habits of trout on Hampshire's River Itchen led to him fishing his flies not on the surface of the water but in the surface film. The technique marked him out as a heretic, but it certainly worked.

A Swarm of Glo Bugs

Even before the weighted nymph debate had begun, the great Glo Bug controversy erupted. The Glo Bug was a simple ball of fluorescent pink wool or synthetic material tied on a caddis hook. The fly, if it could be called such, was said by some to closely resemble trout roe. Its defenders maintained its potency came from its propensity to provoke an aggressive response from trout. Whatever the reason, it proved irresistible to the fish whether fished upstream or down.

According to one source, an American angler living in Tūrangi imported fluorescent yarn[12] and demonstrated its potential with some impressive catches, after which word quickly spread. The Glo Bug appeared in tackle shops and sold in numbers close to 1,000 units per week. In 1984 those fly fishers who took a traditional view of the sport considered the Glo Bug was not a fly at all, and called for the Taupō fishery authorities to ban it.[13] The Glo Bug's supporters countered by pointing to the Red Setter lure with its bright orange chenille body. Was not that lure, long a favourite of the wet fly exponents, a large Glo Bug with a tail, fished downstream?

The authorities refused to impose a ban, probably because it would have been impossible to draft rules defining what was and was not an offending lure. There the matter rested until new regulations emerged at the start of the 2004–05 fishing season. The revised rules, the result of a general review by the Department of Conservation, defined fly fishing as fishing for trout with a fly rod, fly reel, fly line, leader and an artificial fly or flies. They contained no prohibition on the use of flies imitating fish roe, thereby removing scope for any debate over the legality of the Glo Bug or its older cousin the Red Setter. That remains the case today, and the definition of an artificial fly refers only to the materials with which flies are made, such as feathers, fur and wool. There is no reference to colour.[14] By 2015 some anglers were using small pink flies with trailing floss attached in a clear attempt to imitate trout roe with membrane attached.

Benefits and drawbacks of nymph fishing

Nymph fishing had a major benefit for all anglers besides its obvious effectiveness for catching trout. Nymph fishers could cast their flies into parts of pools which were not easily fished with the downstream lure, and thus more fishing water became available for more anglers.

According to a survey in 1990–91, nymph anglers comprised 57 per cent of the fly fishers and took 68.9 per cent of the catch. For the 1995–96 season the corresponding results were 80 per cent of the angling effort and 84 per cent of the total catch.[15] That proportion remained steady, for in 2014 Department of Conservation officials estimated that nymph fishing accounts for 80 per cent of angling effort on the Tongariro River.[16]

However, the system has its disadvantages. A very heavy fly is required to reach trout at the bottom of the Tongariro pools. But with a well-weighted nymph it is difficult to make the long casts necessary to cover water in the wide pools, especially when a long leader is used for depth. Fatigue can set in on the casting arm, leading to poor timing and sometimes an injury as the fly strikes the angler in the back of the head. When hooked with a nymph, the trout often seem flummoxed and the fight with the angler is less exciting; there are no heart-stopping runs down the pool which leave the angler breathless. For those people who want a more relaxed style of fishing, allowing them to feel the electrifying pull as a trout takes the fly, the downstream lure continues to have its attractions.

CHAPTER 21:

Bridges and floods

. . . the Tongariro Bridge will be finished this week. The contractors have made a splendid job of it but as the price was low, they are not much better off for their labour.

— ***THE NEW ZEALAND HERALD*, 25 MAY 1891**

AS THE LARGEST RIVER ENTERING LAKE TAUPŌ, the Tongariro warranted a bridge for traffic from an early stage. In 1891[1] the government erected a one-lane wooden structure 360 feet long. For many years this would be the only bridge on a Taupō river; travellers on the coach road from Tūrangi to Taupō risked a wetting when crossing other streams on the eastern side of the lake.

As the photograph shows, this first bridge crossed the river in a straight line from bank to bank. Floods caused significant damage in the first decade of the twentieth century; one in March 1904, another in late January 1907, then a third the following March. The last of these caused some bridge pilings to be carried away, leaving a gap in the bridge of 30 feet. Wagons from the flax mills at Tokaanu were forced to ford the Tongariro River, and there were several accidents.

After a delay of several years — the government did not seek tenders until June 1911[2] — the repair work was completed in October 1912.[3] For reasons unknown, the repairs were carried out so that the three bridge spans that covered most of the river were set at an angle to the rest.[4] That created not just a gradual bend but a distinct change of direction, and for years afterwards the 'crooked bridge' was notorious among travellers.

The road bridge at Tūrangi, viewed from the eastern (true right) bank of the Tongariro River, 23 January 1898.

Russell Duncan collection, Collection of Hawke's Bay Museums Trust, Ruawharo Tā-ū-rangi, 1320

The Rangipo Prison break

In August 1939 the unusual form of the Tongariro River bridge helped to foil a daring prison break by two inmates from the Rangipo Prison Camp. The men, Bennett and Duff, were members of a work party cutting scrub one afternoon while a warder supervised them. When it started to rain around 3pm the warder, McMillan, ordered the group to take shelter. The men moved off, but Bennett and Duff lagged behind until the others were out of sight, and then overpowered McMillan. After knocking him unconscious they stripped him of his clothes, which Bennett donned in place of his prison garb. The pair then made for the road leading to Tūrangi. Soon after, a light truck drove up, driven by F. J. Carter, a local sawmiller. When the elderly Carter stopped at Bennett's signal, believing him to be an authentic prison officer, the fugitives gave him a choice: surrender the truck or take a beating. Surprised and outnumbered, Carter gave up the truck.

The escapers drove off towards Tūrangi, passed through the settlement and swept onto the Tongariro bridge at high speed. Their plan seemed to be working, but if they were already congratulating themselves, disaster was but seconds away. Perhaps in the euphoria of a successful escape they forgot that the Tongariro bridge was not straight. Whatever the reason, the truck smashed through the side decking and somersaulted into the river, finishing upside down and partly submerged. Duff broke a window and got out, but Bennett remained trapped in the cab. Duff had to work standing in the ice-cold water to free his companion, who was unconscious. By now the alarm had been raised back at the prison camp. The superintendent, Archie Banks, arrived with several warders and took the shivering Duff into custody, but for Bennett it was a trip to hospital to be treated for a suspected fractured skull.

The crash meant the one-lane bridge was closed to traffic for eight hours while repairs were made. This meant delays for travellers, but the affair had one unexpected benefit: there was extra income for Tūrangi fishing lodges from skiers heading for Mt Ruapehu who chose to stay in the village overnight.

Wooden bridge replaced

The wooden bridge was replaced by Wilkins and Davies Ltd in 1955. According to the *Taupo Times*[5] the new bridge (which was made of reinforced concrete) was erected a little downstream from the old one, and is straight, whereas the old bridge was renowned for its kink. It was also built to the width of the main highway, with two lanes instead of one.

The Tongariro River at the main highway bridge has been a favourite fishing location for many years. Perhaps the bridge piles provide a convenient resting spot for trout, and anglers instinctively know this. Whatever the reason, travellers passing through Tūrangi will invariably see at least one fly fisher trying his or her luck in the pool immediately downstream.

Footbridges

For many years the only way to cross to the true right bank of the Tongariro River was by using the main highway bridge at Tūrangi. To reach the pools in the upper reaches meant a very long walk, although

The Bridge Fishing Lodge at Tūrangi, April 1946.
Joyce Galbraith. Supplied by Taupō Museum and Art Gallery.

some people used a small boat to cross above the Island Pool.

In 1935 the Upper Waikato and Tongariro Anglers Club had asked the government for a footbridge to be built near the hatchery camp. The Prisons Department consistently opposed this idea because it owned the land on the far bank and feared that a bridge would be an escape route for inmates.

Some anglers were tempted to cross where the river seemed broad and shallow, but it posed considerable risk. There were two drownings at the Red Hut Pool in the early 1950s alone. One was Sydney Parkes, who had tried to cross with Archie Kitto in a boat hooked onto a wire stretched across the river. Kitto had used this method for years without a problem, but this time it had fatal results. As the boat left the shore on 17 September 1954, Parkes, sitting in the stern, pulled on the rope connecting boat and wire. This caused the vessel to capsize and both men were swept downstream. Searchers recovered Parkes's body from the Cattle Rustlers Pool a couple of hours later.[6]

It was not until 1955 that two footbridges were built over the river above the Major Jones Pool and at the Red Hut Pool to allow anglers easier access to the far bank.[7]

The big flood of 1958 destroyed the Red Hut bridge. It remained out of action for two years until it was replaced by a flying fox. That device remained in place until 1971, when a new bridge took its place. The 1958 flood damaged the bridge above Major Jones, but repairs were made promptly and access for anglers continued as before.

Department of Conservation staff monitor the footbridges periodically to ensure the safety of users, especially after flood events. Occasionally a bridge will be closed temporarily while repairs are made, or the bridge is tested to ensure it will take the intended loads.

Flood events

The 1958 flood, which is covered in Chapter 17, was the largest such event ever to occur on the Tongariro River. But other floods almost as large have been recorded since then. They are a natural part of the river's life. The table below shows the peak flows, measured in cumecs, of floods on the Tongariro river from 1958 to 2022:[8]

Year	Peak flows of floods (cumecs)
1958	1,470
1964	1,037
1967	773
1986	810
1998 (2 July)	814
1998 (9 July)	838
2004 (29 February)	1,420
2022 (19 August)	615

A flood can alter the riverbed but not have a wholly detrimental effect on the fishery. At times the river can also run at a level close to its natural flow. That occurred in the winter of 1993, when high rainfall combined with a maintenance shutdown at the Tokaanu power plant. Wet fly anglers got good results by exploring and finding new locations to fish. Those who fished lies where trout had been taken at times of lower flows struggled to take fish because they failed to adapt to the new conditions.[9]

The July 1998 floods caused ash from volcanic eruptions to wash down the river, making it impossible for trout fry to hatch in the

The old Breakaway Pool, September 1984; a favourite with anglers before it ceased to exist when the river altered its course in 2004.
Eddie Cadogan

river's main stem. However, while the fish population was reduced, the average trout size for the 1998 season was 5.7lb; the best for years.[10] In some areas clean spawning gravel became exposed. The timing was also important, as most trout spawned in September and so the flood damage was minimised.

The 2004 flood was the second largest in Tongariro River history. It saw the Breakaway Pool, a favourite with many anglers, removed completely, but other reaches turned into better pools that were deeper and better defined for the fly fisher.[11] For example, Boulder Reach became good holding water.[12] The 2004 flood also had significant effects on the operations of the National Trout Centre, as described in Chapter 23.

Flooding is a natural part of the river's seasonal cycle. Anglers can only wait to see whether climate change will make this phenomenon become more frequent and of greater intensity.

CHAPTER 22:

Voices for the river

. . . our group is well-placed to take on an advocacy role which looks at twenty-first century problems through twenty-first century thinking . . .

— ADVOCATES FOR THE TONGARIRO RIVER,
ANNUAL REPORT 2002–3

BY THE END OF THE TWENTIETH CENTURY the reputation of the Tongariro River as an internationally renowned trout river was well established. Only two years before, in 1998, it had yielded trout with an average weight of 5.7lb in one of the best winter seasons for many years.[1]

But in fishing nothing is certain, and each season will be different from the one before. In mid-1998 a massive flood had given a reminder that Nature always has the final word. For seven days, 8–15 July, Tūrangi received 195 millimetres of rain; 30 per cent above the normal level. The flood was one of the largest since 1958, causing major damage to river access tracks and bridges.

There were also major forces operating under the surface which, if left unresolved, had the potential to cause a significant decline in the quality of the fishery. These problems stemmed from the changes wrought to the river by the hydro-electricity scheme completed in the 1980s. Tūrangi residents, anglers and local Māori noticed that these had significantly changed the way the river functioned, with reduced water flows causing damage to trout spawning grounds and increasing the risk of flooding on the lower reaches near the delta.

Local ratepayers levied

Two events brought the issue of the Tongariro River's future to a head. In 1999 Environment Waikato, the body with responsibility for administering the region, levied a new type of rate (local tax) on landowners. The aim was to fund the spending required to maintain flood control and river management in the entire Waikato River catchment. Extra taxes are always controversial, so this threat caused Tūrangi people to ask why ratepayers should have to share the cost of river restoration when, had government promises to preserve the fishing been kept, it would never have deteriorated in the first place.

Petition and submission

Then in the following year, 2000, a petition with 600 signatures was presented to Parliament calling attention to the diminished state of the Tongariro fishery. That brought an invitation to make a submission to Parliament's committee on local government and environment issues.

The petition was supported by a submission from Virginia Church, a local lady with family links to the original Māori owners of the land in the Tongariro River delta. The submission, prepared in June 2002, cited the historical significance of the river for local inhabitants, and illustrated the detrimental effects which the power development scheme had caused. The main problem was a gradual build-up of silt in the lower reaches caused by the reduced flow rates required by the scheme. The river carried a heavy volume of silt each year, some 45,000 cubic metres. Under natural flow conditions this material would have been washed into the lake, but it was now accumulating in the river, raising the bed and increasing the flood risk to adjacent land. Excessive willow growth on the riverbanks added to the problem, with the tree roots impeding water flow.

The solution was the removal of rocks, silt and willows. These measures, when combined with increased water flows, would go a long way to correcting the problems. But management changes were also necessary if the river was to be permanently protected, so the submission also urged 'that benchmarks be established which as near as possible represent the state of the Tongariro River before engineering began in the 1960s so that future discussions about the state of this river can proceed effectively'.[2]

Advocates for the Tongariro River formed

A change of government meant that Virginia Church never got the chance to put her case to the politicians in person. But the idea of restoring the river had now gained momentum, and after a series of public meetings a group of Tūrangi people decided to form a society dedicated to the long-term improvement and protection of the Tongariro River. The group first called themselves the Tongariro River Action Group (TRAG) but later changed the name to the Advocates for the Tongariro River (AFTR), becoming an incorporated society in October 2002.

The river had suffered over time because the various bodies dealing with it had differing aims and responsibilities. The Department of Conservation looked after the trout fishery, Environment Waikato was concerned with the waterways in the entire region and the adjacent land, while Genesis Energy sought to produce hydro-electricity from its dams. No one party had overall responsibility for the long-term protection of the Tongariro.

Shortly after it was formed, the AFTR society emphasised this omission in a media release:

> Preservation and enhancement of the Tongariro River are the key objectives of a newly formed action group dedicated to ensuring the well being of New Zealand's most prestigious wild trout fishery.
> Based in Turangi, but canvassing for support throughout New Zealand, the Advocates of the Tongariro River has the primary goal of addressing and promoting conservation issues relating to the management and health of the river and its surrounds.
>
> Acting Chairman, Dr Mark Cosgrove* believes it's important the river has a 'voice' in the face of increasingly complex management issues, which have arisen in recent years.
>
> 'We were concerned that New Zealand's most famous trout river appeared to have had no dedicated group committed to its long term sustainability. Other rivers throughout New Zealand have guardians working to preserve habitats and surrounding environments, but in reality, none of these rivers have the stature of the Tongariro, which is a vitally important wild fishery and economic resource of national importance. We felt it was time to act,' says Dr Cosgrove.
>
> **Mark Cosgrove unfortunately died in 2019. He was replaced as president of the society by retired architect Gary Brown.*

The group's first committee was not made up solely of anglers — although it did include Heather Macdonald, who along with Carol Harwood was one of several female fishing guides. She and the other members had a range of organisational and communication skills, backed by determination, local knowledge and political nous.

The society's vision statement stressed the goal of credibility for the group. It aimed to be recognised 'as an informed, balanced and authoritative group seeking to sustain and enhance the Tongariro river for future generations'.

AFTR's Strategic and Action Plan

At a more practical level, it formed a 10-goal Strategic and Action Plan complete with the steps necessary to carry it out. The goals were:

> To have widespread public awareness of the Tongariro River's past glories, its present state as [affected] by the power schemes and neglect and the potential future degradation if remedial action is not taken.
>
> To ensure the promises made by central and local government and power companies in respect of doing least harm to the Tongariro River are kept.
>
> To establish effective liaison with central and local government, their relevant agencies and with key players.
>
> To gain public involvement and input.
>
> To establish a reasonably objective written statement describing the current state of the river including benchmarks where possible as a basis from which to judge change.
>
> To establish effective liaison with all local groups which have similar interests in the Tongariro River and other rivers with the purpose of co-ordinating lobby and action.
>
> To establish effective dialogue with Iwi and establish mutual trust.
>
> To increase the Advocates' collective knowledge regarding all aspects of the river and of the findings and effective functioning of other like advocacy bodies.
>
> To increase membership.
>
> [To m]aintain financial stability and cash flow to meet planned activity.

The AFTR wanted to argue for better river management from an informed base. That made sound scientific research absolutely essential. Under Dr Mark Cosgrove the society was also determined to make the government accountable by helping to fund the river's restoration after years of neglect. An early media release quoted committee member Heather Macdonald:

> 'While central government has largely abdicated responsibility for river management to local authorities, we believe the government is still a key participant in this river's future. The government's State Owned Enterprise, Genesis, is directly involved in water diversion from the Tongariro for power generation and the river generates millions of dollars in tourism earnings and fishing licence revenue to the benefit of the region and the country. Ultimately, proper management of the river is likely to [be] beyond the resources of local authorities and rate payers, so we believe government needs to be involved,' says Ms Macdonald.

In November 2003, the AFTR held the first Tongariro River Management Forum. This meeting, supported by Environment Waikato, brought together the various interest groups (including angling clubs, local and regional councils and iwi) to link the community and the entities whose activities affected the river. The forum gave ordinary people a chance to influence decision-makers, and in so doing provide a coherent approach which would not merely improve the river but restore it as far as possible to its former state.

As the strategic plan indicated, the AFTR wanted to act on various fronts — scientific, political, historical and local. The 2003 Advocates for the Tongariro River's annual report noted:[3]

> While the Management Forum is not a statutory body, it has the potential to re-establish effective working relationships amongst the agencies and groups which have an interest in the river. Enhancing that relationship is a necessary step in devising an ongoing river management plan.
>
> The background to this development is the abrogation of responsibility by the Crown for the damage incurred as a result of the Tongariro Power Development scheme.

The AFTR would therefore be a catalyst. It would instigate projects to improve the Tongariro for the benefit of the entire community, whether they were anglers, hikers, birdwatchers or tourists, at the same time monitoring the activities of the entities whose activities affected the river.

In March 2004 the AFTR applied political pressure when Mark Cosgrove and other committee members met two government ministers and outlined their concerns over the state of the river. Their basic message was that the Tongariro was a nationally important environmental asset which had suffered years of neglect, and which now needed substantial government funding if it was to be improved. Environment Waikato's proposed Project Watershed, based on adding stopbanks to the lower river, was too little and too late, giving only short-term results.

Cosgrove did not pull his punches, going so far as to call for river management responsibilities to be taken out of the hands of the Department of Conservation and Environment Waikato. His description of the parlous state of the river noted:

> 3. . . . the Tongariro Power Development Scheme (TPD) was financially extravagant, poorly planned, ecologically disastrous and culturally crude. Further, that the principal reason for this state of affairs is the abstraction of about half of the normal natural volume of water to feed the Tokaanu Power Station. This has resulted in abnormal sedimentation and bed raising, channel narrowing, willow proliferation and major changes to the aquatic flora and fauna. Further effects include accelerated flood damage.
>
> . . .
>
> 8. . . . it is time for managers to drop their nineteenth century view of rivers simply as drains and sources of running water for hydro and irrigation purposes and that this obsolete view must be replaced by a twenty-first century view which starts by respecting the ecology of rivers and promotes their sustainability.

The meeting with ministers came shortly after the massive Tongariro River flood of 29 February 2004. The flood underscored the need for urgent river-channel dredging work to allow sediment to travel down to the delta as naturally as possible.

In August 2004, the AFTR followed up with a funding model that proposed shifting more of the financial burden for environmental recovery works from local ratepayers to hydro-electricity generators like Genesis Energy and others who used the Tongariro River for commercial purposes.

The society laid the foundation for sound scientific investigation that year by holding two seminars. The first, given by Department of Conservation technical manager Glenn Maclean, outlined the biology of trout in the Tongariro River, the effects of the power development scheme and results of a fish radio-tracking experiment.

Professor Paul Williams, of Auckland University's School of Environment, gave the second seminar on the Tongariro catchment and how human activities have affected it. Professor Williams also pointed out that the AFTR members were sometimes pursuing conflicting goals when they tried to be both a river management body and a voice for nature conservation. The group's main purpose, he felt, should be clearly identified.

In the years that followed, the AFTR proved itself to be a well-organised group advancing sound ideas for the restoration of the Tongariro River. It formed strong working relationships with the Department of Conservation, Genesis Energy and Environment Waikato, but maintained its independence by giving forceful criticism on matters affecting the river when necessary.

There was also useful dialogue with the National Trout Centre Trust, which operated its museum and aquarium facilities to promote the Taupō trout fishery. The AFTR recognised that it had interests in common with the Trout Centre, and they agreed to co-operate on suitable projects, the first being the creation of riverside walking tracks.

Didymo concerns

The benefits from forming the society were demonstrated in 2005 after the invasive algae *Didymosphenia geminata* (didymo or 'rock snot') was found in New Zealand rivers. First noticed in October 2004, the algae spread across the South Island waterways, most probably being carried between catchments by water birds, trout fishers, kayakers and motor vehicles.

Didymo produced thick brown mats of algae on riverbeds, smothering aquatic life. It caused huge problems for anglers and other recreational river users by fouling tackle, as well as blocking water intakes. By November 2004 it was declared to be an unwanted organism. Law changes made it an offence to knowingly spread the organism, but these were largely ineffective because many transfers between river catchments were made unintentionally.

It took a considerable time for government biosecurity officials to take steps to properly limit didymo, not least because they were trying to fully understand it and the best way of counteracting the threat it posed. It was not until November 2005 that the Department of Conservation launched a public campaign of 'Check, Clean and Dry'.

But the AFTR was determined that the Taupō watershed would not be infected, so in October 2005 the group launched a public education campaign. Some 4,000 pamphlets and leaflets were printed to educate people about the algae and the correct way to prevent it spreading. These were sent to fishing tackle shops and fishing lodges, with posters being placed at many riverside access points.

By early 2006, concerned there was no real action from Biosecurity New Zealand, the AFTR arranged a public meeting to consider ways to prevent didymo incursion. The meeting called for warning signs at domestic airports and ferry terminals, and the use of spray bottles and decontamination stations. Didymo messages were broadcast on the local radio station for six months from mid-2006 (Biosecurity New Zealand assisted with funding for this).

A local motel owner, Ross Baker, displayed a large sign reading **DIDYMO FREE AREA** in central Tūrangi, right where passing travellers would see it. Another public meeting followed, at which it was learnt that felt soles on fishing waders were one way in which didymo could easily be spread. In late October 2007 there was indeed a didymo scare in relation to the Tongariro River, but fortunately it turned out to be a false alarm, as detailed in Chapter 25.

Integrated catchment management plan

A central pillar of the AFTR's work was the drive for an integrated catchment management plan (ICMP). The ICMP concept, advanced by the AFTR in 2007 and supported by the Department of Conservation, local iwi and Genesis Energy, was based on four key fundamentals:

- preventing flood damage to the built environment
- reducing or eliminating biosecurity risks
- facilitating the sustainable production of renewable energy
- keeping the Tongariro River free-flowing.

These key matters were overlaid by strategies designed to implement them, leading in turn to the final single objective of respecting and protecting the life force of the river. The ICMP put the welfare and recovery of the Tongariro at centre-stage. Within the plan each strategy component was linked to the relevant accountable agency. By comparing the current position with the desired objective, the group could measure progress (or the lack of it) and seek funding to make the necessary changes.

When the AFTR advanced the concept of an ICMP in 2007, the various parties involved with the Tongariro River Management Forum, including Environment Waikato and local Māori (as owners of the riverbed), agreed in principle. Implementing the plan would prove to be another story, but obtaining general acceptance from all concerned was an important first step. The big flood of 2004 highlighted the need for such a plan and for the involvement of the AFTR in river management.

The Tongariro River Trail

The AFTR saw the river as a national asset that could be enjoyed by everyone, not just anglers. For that reason the group supported the construction of a riverside track known as the Tongariro River Trail (TRT). The trail, a free amenity, was intended to attract tourists to the district by giving them an easy walk or ride in very pleasant conditions and at any time of the year. It would also provide local employment and take pressure away from the enormously successful Tongariro Alpine Crossing, which was completed by 60,000 hikers each year. One important difference was that the TRT would be open to cyclists.

In 2010 two committee members, Ross Baker and John Wheeler, researched likely locations for new pathways and sought funding from the Taupō District Council for what would ultimately be a three-day walk.[4]

After several years, this community project was completed as a 15-kilometre loop trail giving people of all ages an enjoyable experience that takes in the National Trout Centre and other points of interest in the river's upper reaches. Because the trail follows the river, anglers benefited, too, with more water being opened up to them.

The AFTR's achievements

By 2018 the AFTR could list a number of achievements benefitting the Tongariro River and its surrounds. They included:

- supporting the construction of the Tongariro River Trail
- submitting a Tongariro River Integrated Catchment Management Plan (TRICMP) to the Waikato Regional Council Taupō Zone committee for implementation
- helping to eradicate wilding pines and willow from the river margins
- promoting a programme of planting riverbanks with native plant and shrub species
- campaigning against the proposal to build a carp farm in the Taupō region
- obtaining funding for a post-graduate student to undertake a PhD study of the Lake Taupō food web
- making submissions on the desirable level for Lake Taupō, and proposing an alternative operating scheme
- conducting campaigns to reduce the risk of didymo and other aquatic and plant pests
- maintaining links with other bodies dealing with the Tongariro River
- achieving project value in the order of $500,000.

Trout farming is another issue that the AFTR has had to confront. From time to time politicians and entrepreneurs think New Zealand's rivers and lakes, the heritage of the many, can be taken over for the supposed commercial gain of a few. The AFTR made a submission in 2018 to a parliamentary select committee opposing the idea. In mid-February 2021 the government, well aware that it was looking at a huge controversy, said it would not be acting on the matter for years.

AFTR members engage in a wide range of activities, such as tree planting, track cutting, political lobbying, arranging seminars and watching for and eradicating noxious species. The society keeps the welfare of the Tongariro River as the main focus, with river management carried out on a non-adversarial approach.[5]

The group's record shows what a small group of individuals can achieve in the face of established government interests through determination, thorough research, effective communication, sheer hard work and a willingness to co-operate.

CHAPTER 23:

The Tongariro National Trout Centre

. . . the Tongariro National Trout Centre, from its beginnings as a trout hatchery in 1926 to the present day, has achieved a status unique in New Zealand.

— **JOHN PARSONS, *THE FISHING YEARS***

ONLY TWO MINUTES' DRIVE SOUTH of the town of Tūrangi, a sign informs motorists that they are approaching the entrance to the Tongariro National Trout Centre (TNTC). Turning left off the main highway, the visitor finds a carpark aligned with the road and flanked on the far side by native trees. The Tongariro River is not visible, but a notice at the southern end of the tarmac guides people down a gently sloping path through overhanging shrubbery. After a short walk the path levels out at a small bridge where rushing water is heard. On one side of the bridge is a clear, placid pool on the Waihukahuka Stream, the water so clear and still that it resembles a pond. It is home to huge rainbow trout, which cruise quietly around like submarines. On the downstream side a foaming torrent, the outfall from the pond, gives birth to rapids containing more leviathans, most of them too preoccupied with jostling each other in the prelude to spawning to be aware of human passers-by.

The TNTC is a wonderful facility that anglers and non-anglers alike can enjoy. It is located on the true left bank of the Tongariro River, and occupies the site of the former government trout hatchery. The

Sign near the entrance to the Tongariro National Trout Centre.
W. G. Henderson

hatchery was built in 1926 near the Birch Pool to provide a source of trout fry for stocking rivers and lakes in other parts of New Zealand. Migrating trout were trapped in the Waihukahuka Stream and stripped of their ova, with the fertilised eggs being reared in six hatching boxes fed by water from a local spring.

The 1926 spawning season saw hatchery operations begin under the management of Chief Ranger Frank Yerex. More than 3.4 million ova were collected, probably being sent to the Rotorua hatchery for rearing.[1] The following year the Tongariro River tributaries yielded more than 4 million eggs, with eyed ova being sent to acclimatisation societies within New Zealand and state fisheries bodies in Tasmania and New South Wales.[2] Some 365,000 fry were successfully reared at the hatchery and liberated in Taupō waters.

The Tongariro hatchery still rears fry and fingerlings from Taupō trout, but that is only a minor role. Its main purpose now is to help the TNTC give the public a past, present and future view of the Taupō fishery. When government departments were reorganised by the Labour government in the 1980s, the task of fisheries management

Work at the new Tokaanu hatcheries, Taupo, Waihukahuka Stream, 1927.
Auckland Libraries Heritage Collections AWNS-19270714-42-2

for the Taupō region passed from the Internal Affairs Department's Wildlife Branch to the newly created Department of Conservation. Natural spawning in Taupō rivers has always been so prolific that supplementing the fish population with trout raised by hatchery methods is unnecessary. But officials must also recognise that the fishery is located in a volatile volcanic region. Periodically, eruptions have deposited large amounts of ash in the Tongariro River; the most recent occurring in 1995–96. There is always potential for a lahar (mud spill) to emerge which could decimate the entire Tongariro fishery, so the hatchery must be ready to re-establish the trout population should the need arise.

The Tongariro National Trout Centre developed

The main business of the TNTC today is not trout reproduction but the provision of information and educating the public. Even though natural spawning has always been more than adequate to maintain the trout population, the fishery managers decided to keep the hatchery going but also to make it a visitor attraction. In the early 1980s Lynn Harris, then an officer in the Wildlife Branch of the Department of Internal Affairs, conceived the idea of a national angling museum at Tūrangi.[3] Local people formed the National Trout Centre Trust and worked with the Department of Conservation to provide a virtual shop window for the Taupō trout fishery. Thanks to funding from local businesses and volunteer labour provided by the 20 members of the Turangi Lions Club, a new attraction emerged in the form of an underwater viewing chamber beside the tributary stream running through the hatchery property.[4] Clear-glass panels mounted at the level of the stream bed gave an unrivalled opportunity to see trout activity at close quarters. The chamber quickly became very popular with the public, because it gave the impression of being right among the fish, who were in fact quite undisturbed. A display room also recounted the story of Taupō's trout.

The TNTC became the focal point for the Taupō trout centenary celebrations in mid-April 1983, when a special ceremony was held to commemorate the introduction of rainbow trout to Taupō waters. Politicians, anglers, the media, fisheries managers and the general public watched as a parcel of trout fry was tipped into the hatchery stream in a symbolic re-enactment of that first liberation. Deputy Prime Minister Duncan McIntyre officially opened the building housing the underground trout viewing chamber. It was from this time that TALTAC members began teaching children how to catch trout, in a specially constructed fishing pond.

Up until then, the site had offered an interesting but rather passive experience to the visitor. While there were no entry charges, there was also no one on-hand to outline the history of the hatchery or explain the lifecycle of trout. Information was confined to signs erected at various points. The original concept of a national angling museum had not eventuated because of a lack of funding.

In the late 1980s the Department of Conservation and members of the National Trout Centre Trust, headed by motelier John Milner, decided to make the facility a true National Trout Centre with greater visitor capacity and a more interesting experience. By July 1990, stage one of this project was completed with the construction of a new entrance and carpark. In the years that followed, the Trout Centre, administered by a group of trustees, became one of the district's most popular attractions, particularly with school parties who were taught the rudiments of fly fishing and the importance of stream ecology.

By 1999, annual visitor numbers had reached 50,000 and the facility was charging an entrance fee. This increased size and activity meant the whole operation needed to be put on a more business-like footing.

Working with the National Trout Centre Trust, the Department of Conservation prepared a 10-year plan for further development to make the facility much more informative and educational, at the same time fulfilling an advocacy role for the concept of conservation.[5] The plan (to be completed in stages) envisaged a visitor centre, fixed opening hours, on-site staff to provide information, a salaried teacher to instruct school groups in fishery ecology, extended walking tracks allowing better views of the Tongariro River, an historic angler's hut and smokehouse, the children's fishing pond, an aquarium showing native fish and aquatic plant species, and an improved underwater viewing chamber. There would also be a kiosk beside the pond for volunteer workers to use on children's fishing days, in place of the battered caravan from which they had previously operated.

The TNTC governance structure

In 2000 the governance structure of the operation changed when management passed from the National Trout Centre Trust to a newly formed incorporated society, the Tongariro National Trout Centre. The society's committee, led by chairman John Milner, operated the Trout Centre under a lease from the Department of Conservation and owned the various buildings (visitor centre, shelters, workshops and so on).

The society committee worked hard to source funding from businesses, charitable foundations and individuals, and then started to make the concept plan a reality. The River Walk is the name the society committee gave to the visitor centre housing historic fly fishing tackle, a fly tier's hut, an aquarium, a library and other informative displays. Designed by Peter Langford, it was erected in 2003 at a cost of $500,000. The building's name implies the experience of a Tongariro angler making his or her way along the river, anticipating new and exciting encounters around every corner.

The weather gods were kind when Prime Minister Helen Clark officially opened the building on 28 August 2003 in bright sunshine. Some 150 guests attended the opening, with speeches being made by representatives from Ngāti Tūwharetoa, the Department of Conservation, Genesis Power Ltd, the Mayor of Taupō, and John Milner. The prime minister also viewed the children's fishing pond, where future anglers from the local primary schools were tempting the trout.

The 2004 flood

February seems to be the time for high-water events in the Tongariro River. The big flood of 1958 occurred in that month, and the pattern was repeated in 2004 when 133 millimetres of rain fell in Tūrangi on 29 February. The deluge made the Tongariro River break its banks to the south of the Trout Centre, swamping the trout rearing ponds and entering the River Walk building. All the juvenile trout in the hatchery building nearby were destroyed, while their larger 18-month-old cousins in the fishing pond were swept away, at risk of being killed by suspended silt clogging their gills. Water reached a depth of 400 millimetres in the adjacent kiosk.

While the damage was extensive, it was not completely irreparable. Local people repaired much of it with the help of the Turangi Volunteer Fire Brigade, by clearing the flood debris, cleaning walking tracks and removing mud from the rearing ponds. The Trout Centre had to be closed for a period, but re-opened on 10 April 2004, although the loss of trout meant children's fishing days had to be cancelled for the winter. Surprisingly, a Department of Conservation fisheries survey showed that juvenile trout numbers in the river were good, while some of the trout that had inhabited the fishing pond appeared in the Waihukahuka stream apparently unscathed, and joined the wild trout there.

The flood did have some other positive results. It created new pools, with some shallower stretches that promised to be good habitat for the growth of young trout resulting from the winter's spawning.

The education programme begins

To drive home the importance of conservation to future generations, the TNTC society started a Taupō for Tomorrow education programme in 2003. The programme is a joint venture between the Tongariro National Trout Centre, Genesis Energy and the Department of Conservation. The Whakapumautanga Downs Learning Centre, located beside the children's fishing pond, is named after a local Māori man, now deceased, who was a patron of the society. Officially opened in April 2006, the learning centre has a full-time education officer, who can instruct visitors (both students and adults) about freshwater ecology, the management of natural resources, sustainable energy and the art of fly fishing.

The instruction provided by the Learning Centre is spread to schools throughout New Zealand thanks to the LEARNZ virtual-reality programme — an online education facility that enables students to experience the Taupō trout fishery from afar. A notable example occurred in August 2007, when 52 groups from different schools took part in a three-day virtual field trip around the fishery, studying trout biology, fish trapping and tagging, pest prevention measures, a visit to a spawning stream, recreational fishing and law enforcement. Audio conferencing enabled students to ask questions of rangers from the Department of Conservation, based on the course study materials which had been distributed in advance.

Aquarium and other developments

In the years that followed, the Trout Centre was such a drawcard for tourists that improvements had to be made to the main building. Plans to install the aquarium were drawn up and approved in 2007, a fish-pass display constructed, and the riverside paths extended. All of these changes were funded largely with sponsorship contributions from corporations, Genesis Energy, the Department of Conservation and private organisations such as the Sargood Trust. The 20 years from 1987 to 2007 saw the TNTC and its predecessor trust raise $952,000 for investment in new and better facilities at the centre.[6]

Nor has the original concept of an angling museum been overlooked. Greenheart and split-cane rods from bygone days hang on the centre's walls, while vintage fly reels and minnows from such venerable firms as the House of Hardy are displayed in glass cases. The River Walk Visitor Centre received a new entrance and reception area in December 2009, with public toilets installed close by. The new entrance linked to the public carpark; an important feature, because feedback from visitors showed many of them got lost on the paths threading through the native bush near the river's edge and could find neither the Visitor Centre nor the toilets. For the volunteers, essential to the smooth running of the centre, there was a dedicated work room.

After years of careful planning by the TNTC committee, Genesis Energy and the Department of Conservation, the aquarium officially opened in December 2011. Kōura (freshwater crayfish), bullies, inanga, kōkopu, kōaro, mudfish and eels are among the native fish swimming in glass tanks lining the walls of the display. Eels apart, in the wild these species go unnoticed by many people, and so the aquarium has an important role in educating visitors about the importance of protecting them.

The fish are shown in settings replicating their normal aquatic environments, such as ponds, marshes and streams. Undesirable pest fish (koi carp and catfish), which provide a threat to the nation's ecosystems, are also shown.

Volunteers give invaluable assistance to the TNTC, which could not function without them. They help run the Trout Centre tours, staff the River Walk museum, which is open for all but two days in the year, maintain the grounds, instruct children at the fishing pond, and generally deal with any problems arising. For example, when the TNTC decided to offer to smoke any trout the children caught in the fishing pond, the demand for this service quickly caught on. TNTC volunteers, who were using small cooking equipment designed to take one or two trout, found themselves unable to keep up with the demand. So John Milner decided to install a large trout smoker to process the fish in bulk, which Gianni Salvador built in line with John's design. It proved to be very successful, with its large gas ring enabling the smoker to deal with several dozen trout at once.[7] The TNTC recognised John Milner's many years of support by making him its first life member, in August 2004.

The Tongariro in prose and paint

In 1998 Taupō artist Val Raymond contributed to the rainbow trout centenary celebrations with an exhibition of paintings called *Tongariro River Heritage*. The exhibition, held at the Taupō Museum, comprised five groups of 127 paintings tracing the river from its birth on Mt Ruapehu down to the delta.

She was fascinated by the landscape and flora of the central plateau region, and it forms a central theme of her work. The Tongariro National Trout Centre holds a collection of paintings she gifted to the organisation.

Val Raymond died in 2021 aged 89 years, having made an important contribution to the arts scene nationally and to the history of Taupō.

In June 2019 there was a special function at the TNTC to mark the life of the late John Parsons, author of several books recounting the delights of angling on the Tongariro river and other streams so dear to his heart. He also researched trout fishing history of the Waikato

River, culminating in the book *Pye's Kingdom of Huka Lodge*.[8] John Parsons never sought to formally instruct others in the art of fly fishing; he simply enjoyed writing about it — and he wrote very well. Even so, there is much practical knowledge that anglers can glean from his work, as he shares the triumphs and disasters (both large and small) of each outing, ponders why this or that fly succeeded or failed, considers Lake Taupō's dragonflies, its bird and other animal life, and anything else he encounters while practising the gentle art.

The Trout Centre's success and benefits

By 2016 the River Walk centre was receiving 70,000 visitors per year. It is one of the top three tourist attractions in the Taupō region, showing people how the fishery functions and educating future generations about the importance of water resources. The opening hours (during which an access fee is charged) are from 10am to 3pm, but the facility can be used at other times and in a variety of ways which reflect the Taupō fishery and the anglers who use it.

While trout are the main focus of the TNTC, visitors may get to see rare native wildlife in the form of the blue duck (whio). The TNTC operates a rearing project for young birds, which aged around three months are collected outside the Tongariro region, before being returned to their home river.

The TNTC does not own the land on which it is located, but instead leases it. In 2017, part of the property was transferred to Ngāti Tūwharetoa under a Treaty of Waitangi settlement. The settlement saw many parcels of land in and around Taupō transferred by the Crown as redress for historical breaches. Ngāti Tūwharetoa also gained representation on the body administering the centre.

The society and the iwi negotiated a new lease at a peppercorn rent, and the Trout Centre began to be operated jointly by trustees representing the society, the iwi and the Department of Conservation. To all appearances things carried on as before, with children's fishing days scheduled regularly, visits by school parties hosted, and other educational activities provided.

The undoubted success of the Trout Centre is reflected in the figures for trading income recorded in the annual financial statements. For the 2009/10 year income earned was $5,216, while for the 2018/19 it was $215,061.

The TNTC does more than provide a superb opportunity to learn about the art of trout fishing. It also sets out the history of the Taupō fishery and provides a unique insight into the ecological web that is integral to the trout's existence. The displays are detailed and informative, and there are knowledgeable people on hand who can answer any questions visitors may have. Best of all, the centre can be visited repeatedly because with the change of seasons a new scene presents itself.

CHAPTER 24:

Trout by numbers

During the early years of the Taupo fishery, the size of the trout population was thought to be so huge that the overall catch by anglers would be relatively insignificant.

— *TARGET TAUPŌ*, ISSUE 64, JANUARY 2012

IT IS SAID THAT AN INVESTOR once asked the United States financier J. P. Morgan what the share market would do. Morgan replied acidly that the market would fluctuate. The same applies to trout fishing seasons on the Tongariro River — each one differs from the last. The variation results from any number of factors, ranging from altered climatic or water conditions to angler pressure.

The Taupō trout fishery, and with it the Tongariro River, has been studied by scientists for more than 50 years. The lessons learnt from these surveys are applied to ensure the continuance of environmental conditions fostering 'a huge, self-sustaining population of quick growing, superbly conditioned trout providing high catch rates of acceptable sized fish'.[1] This goal is achieved by using management tools such as angler satisfaction surveys, fish counts, environmental studies and bag and size limits.

Fishery statistics

Records taken by the Department of Conservation tell the fishery's story. The graph below tracks the average weight of Taupō trout caught from 1900 to 2010, with major peaks in 1910 and the mid-1920s.

Target Taupō, July 2011, Issue 63, p 27

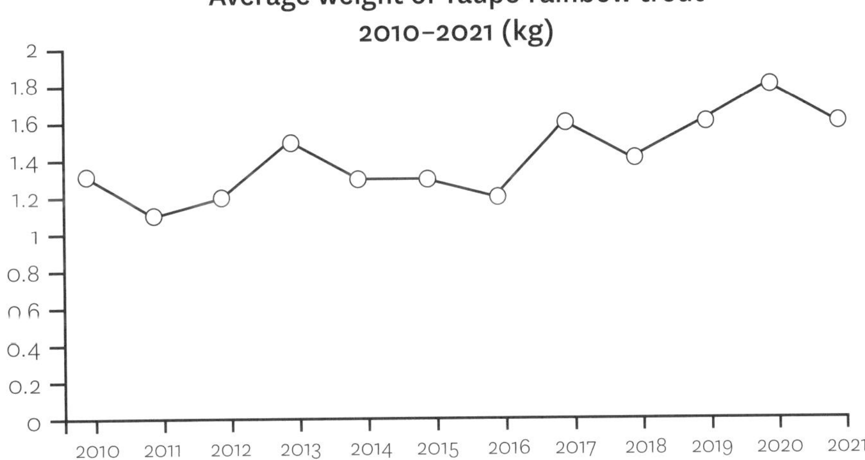

Note the statistics are not restricted to trout from the Tongariro River.

Department of Conservation

Variability in trout weight has occurred at other times. In 1994 Taupō endured one of its wettest winters. In the years preceding, the trout checked in the Whitikau Stream trap had averaged 1.8–1.9kg (3.9–4.18lb). In the first half of 1994 the average weight was more than 2kg (4.4lb).[2]

Fisheries officers, commenting on early season fishing for 1994, noted that the trout were larger and stronger:

> We received numerous comments from anglers about how difficult it was to land some of the fish hooked and the frequent occurrence of fish passing through the [Whitikau Stream] trap with flies in their mouths bore testimony to this.[3]

It is noticeable that for the 20 years from 1985 to 2005, the average weight of trout caught exceeded the average for the period from 1955 to 1985.

The trout population also swings wildly. In the eight years from 1998 to 2006, using the fish trap on the Waipa Stream as an indicator, the number of trout entering the Tongariro River increased from around 2,000 to almost 6,000.[4] By 2012 the number was back to 3,000. The variation in fish numbers corresponds to the three-fold change in numbers of large trout in Lake Taupō as recorded by acoustic surveys.[5]

Angler numbers and fish harvest surveys

The number of fishing licences sold each year indicates the level of angling activity. Presumably, almost every person who buys a fishing licence actually uses it. However, licences sold do not equate to the number of anglers at Taupō, as the same person may buy a day or month licence several times in the course of a season. It appears that the state of the economy and improvements in angler catch rates are two important factors governing angler numbers.

In the 50 years spanned by the 1950/51–2000/01 seasons sales of fishing licences at Taupō show an overall increase, beginning at 10,000 and peaking at more than 82,000 in the late 1980s. A general decrease in licence sales then began. At first it was gradual, but a sharp decline occurred in the first decade of the 21st century. Sales in 2000/01 of 74,293 fell to 41,363 in the 2011/12 season. In broad terms, licence sales for 2011/12 were around half the number for 1987/88. Numbers continued to decline and by 2014 total licence sales were 39,513.

Various social and economic factors influenced anglers on whether to fish at Taupō. A 2017 study co-authored by Department of Conservation scientist Michel Dedual concluded that two key elements affecting the total licence sales to population ratio were unemployment and the quality of the fishery. In terms of angler numbers it seems the Taupō fishery has yet to reach full capacity.

Another statistic shows the trend in recent years as to the number of fish actually caught. *Target Taupō* recorded the results of the 2011 harvest survey conducted every five years on Lake Taupō, and noted:

> The percentage of legal-size fish harvested on the Tongariro river since season 1990/91 has steadily declined since season 1990/91 from 75.5% 20 years ago to 36.3% during season 2010/11. This could be due to a

> number of factors including the size and quality of fish during more recent years and the timing of the spawning runs coming much later.[6]

In consequence, fishery managers considered the need to reduce the minimum size limit for takeable trout to 35 centimetres. That reduction first applied to the season beginning on 1 July 2017. Variation in the average weight of Taupō rainbow trout is demonstrated by Waipa trap records for 1998–2001.

1998	1999	2000	2001
2.4kg	1.72kg	1.95kg	1.88kg

The 1998 season was exceptional for the size and quality of trout taken. However, in 2008 fishery managers encountered the poorest fishing season on record. This decline was attributed to settled weather conditions in the winter of 2005, which meant that the water strata in Lake Taupō failed to mix, preventing essential nutrients rising from the lake bed. These conditions had the effect of denying smelt the zooplankton they required, breaking the food chain so that smelt failed to materialise in their usual numbers. As a result, Taupō trout had a very restricted diet with minimal or non-existent growth rates.

Fortunately, the fishery recovered from this setback, and in following seasons anglers were once more catching trout of respectable proportions.[7] In 2019, the average weight of female and male rainbow trout caught in the Waipa Stream trap for the months May to October was 3.5lb (1.6kg). For 2020, rainbows had increased in size and length with an average weight of 3.9lb (1.8kg).[8]

Michel Dedual, Department of Conservation fisheries scientist, concludes:

> We should certainly cherish the wild aspects of the Taupo fishery which is the envy of the rest of the world. The way by which the fishery adjusts and bounces back following changes in environmental conditions is simply incredible and we should have faith that trout will not disappear. After all they have been around for more than 100 million years and survived the disappearance of dinosaurs. Our job as fishery managers is to hopefully achieve a balance, making sure good water quality and quantity persist in the Taupo catchment but at the same time have good numbers of quality fish available.[9]

CHAPTER 25:

The future of the fishery

The future depends on what you do today.
— **MAHATMA GANDHI**

THE TONGARIRO RIVER is the largest river entering Lake Taupō, but as we have seen it is only one small part of a much larger catchment. The entire watershed must be maintained in a healthy state if the trout fishery is to endure for the benefit of future generations. There is little point in protecting and nurturing the riverine habitat if Lake Taupō has such poor water quality or inadequate food supplies that trout cannot thrive there. The reproductive cycle would be impeded, leading ultimately to inferior fish in the angler's bag.

Lake Taupō and its rivers have long been recognised as important environmental assets for New Zealand. However, there have been major changes to the physical environment in the latter half of the twentieth century. In 1941 the installation of control gates on the Waikato River outlet at Taupō town raised the lake level several feet. It remained at that level for far longer periods than would have occurred naturally. In the mid-1960s work began on harnessing the Tongariro River for hydro-electricity purposes, reducing its flow and altering the normal pattern of seasonal freshes.

There have also been changes in land use, with the tussock and native vegetation on the region's pumice soils replaced by production forest and farm pasture in some areas. Historically, Lake Taupō had low levels of plant nutrients, but once farming and forestry began increased amounts of nitrogen and sediment entered the lake, reducing water quality by encouraging the expansion of algae and phytoplankton.

Watershed preservation

By the 1960s, Taupō local authorities had become concerned about possible threats to the lake and its rivers. A system of lakeshore reserves was established in an effort to limit the effects of fertiliser runoff from adjacent farms. Catchment protection projects followed in the 1980s, which saw tree planting and fencing on the margins of the rivers and streams feeding the lake.

Official reports issued in 1998 and 1999 indicated that water quality in Lake Taupō had declined. The Waikato Regional Council, which had a statutory obligation to protect the lake, could see more dairy farming looming and so decided to impose new rules limiting the ways in which farmland could be worked in the Taupō basin. It was the first time that non-point-source pollution would be managed in that way, and the farmers affected were greatly concerned at the potential loss of income and reduction in property values. After years of meetings and negotiation, the local body administrators, farmers and other interest groups reached a common position on how to deal with the pollution problem.

In 2003 the New Zealand government agreed to commit funds to a programme to maintain the lake's water quality by gradually switching nitrogen-leaching pastoral land surrounding Lake Taupo to low-nitrogen land uses. The programme was implemented by the Lake Taupō Protection Trust, a charitable body officially launched in February 2007 with the aim of permanently reducing the amount of nitrogen entering the lake by 20 per cent no later than 2020. The trust achieved its 20 per cent target in 2015, largely through purchasing land and converting it to low-nitrogen uses. It also set up the Lake Taupō nitrogen market which sees non-point sources of nitrogen operate under a cap. In 2021, government involvement ceased and the trust continued to operate with funding from Environment Waikato and the Taupō District Council.

While the nitrogen reduction scheme met with a degree of success, some sources of this element, such as forestry and the Tongariro hydro-electricity scheme, cannot be eliminated and problems still remain. The potential for a decline in Taupō's water quality was demonstrated in December 2017[1] when algal blooms appeared in five places around the lake, forcing the cancellation of the swimming leg of the Taupō Ironman athletic event. The presence of potentially toxic algae in New Zealand's largest lake prompted a public health warning from the authorities. It was a poor advertisement for the tourism sector, which for years had promoted a '100% Pure New Zealand' experience to the overseas market.

In 2007, the Crown and the Tūwharetoa Trust Board signed a deed recognising the trust board as owner of the bed of Lake Taupō and certain rivers entering the lake. The deed records an agreement that these 'Taupō waters' are to be administered as a reserve under the Reserves Act 1977 by the Taupō-Nui-a-Tia Management Board.

On 31 July 2020 the Taupō-Nui-a-Tia Management Board issued a draft management plan for Taupō waters and invited public submissions. In broad terms the plan aims to manage Lake Taupō and its rivers so that non-commercial public access (such as for recreation) continues while environmental values are protected.

Threats to Lake Taupō Fishery

The Department of Conservation manages the Taupō trout fishery on behalf of the Crown and in conjunction with Ngāti Tūwharetoa. The task of fishery managers is to provide the maximum number of anglers with the best possible wild trout fishery. In the twenty-first century they face an array of potential threats, including:

- pollution
- pest species (both fish and aquatic weeds)
- damaging human activities
- the effects of climate change.

Pollution

Because the Tongariro River rises in a volcanic region mostly covered with native vegetation, water quality is high and direct contamination is negligible. However, from time to time fisheries managers have had to counter pollution threats. In December 1957, the Conservator of Wildlife learnt that the Rangipo prison farm was planning to erect a new milking shed and discharge 400 gallons of wastewater per day into the Tongariro River. Even though the environmental effects would have been minimal, Pat Burstall, the senior fisheries officer, strongly opposed the idea.[2]

As we have seen, the Tongariro National Trout Centre, located on the upper river, attracts thousands of visitors annually, which ironically could pose a threat to the river. So to protect water quality, the centre's toilet facilities are linked to the Tūrangi sewerage system.

Pest species

The only pest fish to appear in Lake Taupō in recent years has been the brown bullhead catfish, thought to have been introduced in 1985. This species competes with trout for food, although it does not appear to have detrimentally affected the fishery. Catfish do not seem to be present in the Tongariro River.

The invasive weed *Didymosphenia geminata* (didymo or 'rock snot') is a potential problem for Taupō rivers as detailed in Chapter 22. In late October 2007 there was a major scare when river monitoring revealed dead didymo cells were present in the Tongariro, Whanganui and Whakapapa rivers. The discovery forced Genesis Energy to cease diverting water across the Tongariro power scheme in order to limit the spread of the algae. Fortunately, the incident was a false alarm. The dead didymo had been contained in sample bottle caps sent from the South Island.[3] The one positive outcome was an opportunity for Department of Conservation staff to apply their pest containment strategies.

Damaging human activities

A range of human activities have been identified as posing possible threats to the desired water quality of Taupō waters and the health of the fishery, including trout farming, land development, motor vehicles, anglers' use of nylon leaders, as well as the advancing threat of climate change.

Trout farming

A major issue arose in mid-2015 when news broke that the Office of Treaty Settlements, acting for the Crown, had agreed that Ngāti Tūwharetoa would have the right to use a race at the Tongariro National Trout Centre (TNTC) to raise trout for ceremonial and other cultural purposes. Fish and Game New Zealand, which has a statutory obligation to protect and foster the nation's trout fisheries, became concerned the agreement opened the door to a trout farming operation. However, a spokesman for the Tūwharetoa Hapū Forum was reported as saying that negotiations were still in train and commercial fish farming was not planned.[4] The formal deed of settlement between the iwi and the Crown in 2017 supported that statement with a clear reference to rearing fish for cultural purposes only.[5]

In May 2018 a former salmon farmer, Clive Barker, filed a petition at Parliament seeking a change in the law prohibiting trout farming. The petition drew support from commercial organisations and iwi groups, including the Tūwharetoa Māori Trust board (although the latter expressed the need for a robust assessment of biosecurity risks against purported economic benefits). Recreational fishing bodies opposed the petition.

In 2019 the Lake Rotoaira Forest Trust, a Ngāti Tūwharetoa entity, was reported as viewing Lake Rotoaira as a suitable site for aquaculture. Other reports noted the trust considered that the law preventing the farming and sale of trout in New Zealand should be repealed. It appears that the trust's objective was to open the way for a trout- and salmon-rearing operation near Mt Tongariro. Details were sketchy, but indicated that fish were to be raised in tanks, with waste applied to adjacent land on which there would be an allied horticultural venture.

The prospect of trout farming in New Zealand has long generated opposition from recreational anglers. However, the idea seems to have a fatal attraction for politicians, who invariably delude themselves that trout farming will create many jobs, generate huge export earnings and garner votes. In the 1972 general election it was one of several controversial issues which helped to unseat the National government.

Experience in other countries shows that the threats trout farming poses are real and many, including the spread of fish diseases, depletion of the wild trout gene pool, and an increased risk of trout poaching. Water pollution is another threat, as a trout farm is simply a feedlot where the fish are massed in pens, leading to high concentrations of fish waste in the surrounding waters. As Lake Rotoaira is the source of the Poutu River, a major Tongariro tributary, if any waste products from fish rearing were to reach that water body they would eventually end up in Lake Taupō.

International trade is another point often overlooked. Were New Zealand to licence trout farming, its obligations as a member of the World Trade Organization would require it to allow imports of farmed fish from countries like China. Because overseas fish farms are very large operations, it is highly likely that New Zealand producers could not hope to compete on price.

Commercial fish farming has established itself in Australia, but has proven controversial and sometimes environmentally damaging. In 2018, Huon Aquaculture, a Tasmanian salmon farming company, received the award of Australian Farmer of the Year. Two years later Huon Aquaculture received less welcome publicity when it pleaded guilty to six counts of breaching environmental regulations after illegally dumping 80,000 litres of wastewater.[6] The company's managers claimed to be ignorant of environmental protection notices issued to prevent such offending.

To recreational anglers it makes no sense to gamble on a venture with uncertain benefits and the potential for significant problems. Taupō is not just a superb wild trout fishery with an international reputation, it is also economically important for the region and provides some 300 jobs.

In September 2020, Parliament's select committee on primary production, which seemed wilfully blind to the lessons from history, recommended that the government give serious consideration to commercialising trout farming. However, in February 2021 the government said it would not be reviewing the law because no analysis had been done on trout farming's supposed benefits, nor on the risks to New Zealand's wild fisheries.

Land development

Human habitation and industrial activity inevitably result in deposition of rubbish and detritus in the environment, either deliberately or through neglect. The restrictions on the types of land use permitted in the Taupō basin are intended to limit such problems.

In 2008 a joint venture company announced a scheme to develop a small town on 650 hectares of farmland near Tūrangi. Some shares in the company were held by a company owned by two businessmen, while the balance was held by the trustees of a trust representing Māori in the Taupō district.

The land, which had belonged to a government department, had been removed from production to reduce nitrogen leaching into Lake Taupō. The new plan proposed the construction of 2,500 house lots, a golf course, a fishing lodge and two hotels. Conservation group Advocates for the Tongariro River was concerned that the development would take place near the Mangamawhitiwhiti Stream, an important brown trout spawning site.

The town project did not get very far, possibly because it was conceived at the time of the global financial crisis. The joint venture company borrowed funds and bought the land, but difficulties arose when the scheme's financiers brought in another lender and restructured the loan. In legal terms that meant the company's shareholders became liable for the debt, effectively guaranteeing it. The trustees of several other Māori trusts — none of them with experience in commercial land deals — were also pressured into signing loan documents without receiving independent advice from a lawyer. As a result some of them inadvertently took on liability for a large debt. After the company defaulted on its loan it went into receivership in 2009, leaving some of the trustees who had committed to the loan facing a debt of more than $4 million.[7]

Motor vehicles

Motor vehicles are another form of human activity that can affect the catchment. Since 1922 motor cars have travelled from Tūrangi to Taupō on a road built along Lake Taupō's eastern shore. The road, which began as little more than a track carved out of the pumice soil for cars like the Ford Model T, must now accommodate heavy articulated trucks.

The highway (State Highway 1) follows the contour of the lake, often passing a stone's throw from the water. In several places it rounds a hairpin bend where a rocky point juts into the lake. Here the road is elevated. Large transport vehicles find it hard to negotiate the narrow road and can sometimes end up in the water. In 1998 a truck-and-trailer unit carrying frozen foods plunged down a steep bank and into the lake,[8] closing the road to traffic for most of the next day. A similar accident in 2014 saw 700 litres of diesel fuel spilled into the lake.

Motor vehicles in transit deposit a number of contaminants on the highway, including copper, zinc, oil, fuel and grease. These elements find their way into the lake as stormwater runoff, and have the potential to alter the water chemistry. While the Taupō District Council is aware of the problems posed by stormwater, its policy is directed more to runoff from urban areas.

Angling

Even the gentle art of angling can affect waterways adversely. Since nylon replaced silkworm gut as a leader material in the years after

World War II, huge amounts of it must have been left on rocks and snags in Taupō's rivers and in the surrounding vegetation. Nylon takes hundreds of years to degrade (fluorocarbon takes even longer), and in that time it can be a trap for water birds and other wildlife, as well as diminishing the quality of the angling experience. In future people may be required to fish with biodegradable leader material.

Climate change

Rising temperatures are one of the effects of climate change, and have the potential to alter the entire range of New Zealand's ecosystems. The likely consequences are thought to be rising sea levels, more extreme weather events (droughts and floods), and changes in rainfall patterns across the country.

The New Zealand government enacted the Climate Change Response (Zero Carbon) Amendment Act 2019 to establish and implement climate change policies that contribute to the global effort, under the Paris Agreement, to limit the global average temperature increase to 1.5° Celsius above pre-industrial levels.

It is impossible to make an accurate forecast of how climate change will affect the Tongariro River fishery, but extreme weather events must surely be detrimental in the long term. A 2018 study by the National Institute of Water and Atmospheric Research (NIWA) used scientific modelling to estimate the likely effect of climate change on New Zealand rivers.[9] For North Island rivers the projections were that by the late twenty-first century there would be reduced annual and mean low flows. The Waikato would possibly have an increased mean annual flood figure and more frequent floods of three or more times the median flow. Because the study was based on river mouths, its results cannot be extrapolated across entire catchments and further studies will be essential.

Another study by officials at the Ministry for the Environment concluded climate change would mean reduced snowfall, with a consequent effect on flows in rivers such as the Tongariro that rise in alpine regions.[10] In January 2021 the international journal *Nature* published the results of a study into the effects of higher water temperatures on lakes worldwide. The study foresaw major problems for freshwater marine life from thermal changes in lakes, and concluded darkly that in the latter twenty-first century species and their ecosystems might be pushed to the limits of survival.[11]

Extreme weather events occur at Tūrangi occasionally, as shown by this snowfall on the Tongariro River.
Copyright estate of John Parsons

In 2002 Department of Conservation scientist Michel Dedual discussed the potential effects of increased temperatures on Lake Taupō. The increased volatility of river flows and water temperatures could have adverse results for the trout fishery by altering trout habitat and the invertebrates on which trout feed. He also noted that lakes are extremely vulnerable. In the end, he concluded that just as humans are causing climate change, so too they can act to find a solution.[12]

Stewardship required

Despite all that has happened in more than a century, the Tongariro River is still an outstanding trout fishery attracting fly fishers from all over the world. All of those who enjoy it are charged with the responsibility of protecting it and thereby ensuring that the quality of the fishing is maintained for future generations of anglers.

Appendices

Appendix A: Agreement between the Crown and Ngāti Tūwharetoa on Taupō waters, 1926

OFFICE OF THE MINISTER OF NATIVE AFFAIRS,
WELLINGTON,
26th July, 1926.

MEMORANDUM for

Right Hon. J. G. Coates,
Native Minister,
WELLINGTON.

Fishing Rights in Taupo Waters.

I have the honour to report that following on the terms of settlement agreedto between yourself and Ngati-Tuwharetoa last Friday the Hon. Mr. Ngata, Mr. G. P. Newton, and I drew up the following agreement. The agreement was translated into the Maori language and was read out to the representatives of Ngati-Tuwharetoa, who directed Hoani Te Heuheu to sign it on their behalf. The two copies in English were so signed, as well as the two copies in Maori--two for the Crown and two for Ngati-Tuwharetoa.

These are respectfully submitted for your consideration and signature, if you approve.

If you approve, the necessary legislation will be drafted immediately for inclusion in this year's Native Washing-up Bill

Private Secretary.

Enclosure.

Taupo Waters and Fishing Rights.

THE TERMS discussed and provisionally agreed upon at a Conference between the Prime Minister and the Ngati-Tuwharetoa Tribe held at Waihi, Tokaanu, on the 21st day of April, 1926, relative to Taupo Waters and fishing rights therein were further discussed in Wellington on the 23rd day of July, 1926, when the Prime Minister met selected representatives of the said Tribe:

A final agreement was arrived at upon all main points at issue and the Ngati-Tuwharetoa representatives nominated Hoani Te Heuheu to sign the agreement on behalf of the Tribe.

The main points of the agreement are as follows:

1. The term 'Taupo Waters' has the meaning defined in section 29 of the Native Land Amendment and Native Land Claims Adjustment Act, 1924, namely, that it means and includes Lake Taupo and all rivers and streams flowing into that Lake, and the Waikato River between Lake Taupo and the Huka Falls.

2. The Government shall, subject to the provisions of the next succeeding clause, pay to a Board hereinafter constituted the sum of £3,000 a year to be administered for the general benefit of the members of the Ngati-Tuwharetoa Tribe or their descendants.

3. Should the amount of fishing license fees, rents of Camp sites and revenues hereinafter referred to derived from Taupo Waters and lands over which access thereto is reserved under this agreement exceed the sum of £3,000, then one-half of the amount of the excess shall in like manner be paid to the said Board for the purposes mentioned.

2.

4. The beds of all Taupo Waters shall be vested in the King as a Public Reserve.

5. The public shall have access to and right of passage over a margin of one chain from and around the bed of Lake Taupo.

6. Every holder of a special license shall have access to and right of passage by foot over continuous strips of land one chain wide on each side of and such length of the beds of such Taupo Waters other than the Lake as may from time to time be defined by Order in Council.

7. Fences erected or to be erected upon or across lands adjoining the beds of Taupo Waters shall not be deemed to obstruct any right of access or passage if reasonable means of passage by gates or turnstiles are provided through or over such fences.

8. On the recommendation of the said Board certain areas within the one chain strip may be excluded from public access by Order in Council.

9. Power shall be given to the Minister of Internal Affairs to permit and control the erection of fishing camps on the chain wide strip and to levy and collect rent therefor, such rent being deemed revenue within the meaning of Clause 3 hereof.

10. The Minister of Internal Affairs shall be empowered to license at a fee to be prescribed by regulations all boats or launches plying for hire on Taupo Waters.

11. One-half of the amount of fines imposed for the breach of the Fisheries Act shall be deemed to be revenue within the meaning of Clause 3 hereof.

3.

12. Fifty free licenses shall be granted to members of the Ngati-Tuwharetoa Tribe to be nominated by the said Board.

13. Whereas the legally ascertained owners of lands bordering on certain streams flowing into Lake Taupo or persons claiming through them allege that in the past they have derived substantial revenues from the use of their lands as Camp sites for anglers or for fishing purposes; and whereas it is desirable that such claims and allegations shall be investigated with a view to their settlement by payment of monetary compensation or otherwise, it is hereby agreed that the Government shall appoint a tribunal to investigate such claims and allegations and to determine what compensation, if any, shall be given in settlement of the same.

14. The Board hereinbefore referred to shall be called the Tuwharetoa Trust Board; shall consist of eight members, being members of the Ngati-Tuwharetoa Tribe, appointed by the Native Minister after receiving nominations or recommendations from the Ngati-Tuwharetoa Tribe. The term of office shall be two years. Regulations shall be prescribed providing, inter alia, for details of administration, payment of members and their expenses and for the Board's account.

15. Legislation shall be introduced this session to give effect to this agreement, and shall, inter alia, include such provisions of the legislation relating to the Arawa Lakes as may be applicable.

Witness to signatures — *Balneavis, Private Secretary, Wellington*

Hoani te Heuheu

Appendix B: The 'Shand agreement': the 1964 letter from the Marine Department to the Taranaki Acclimatisation Society concerning safeguards for the Tongariro River fishery

Mr. Brian Quickfall
Secretary,
Taranaki Acclimatisation Society,
Box 57, New Plymouth

Dear Mr. Quickfall,
The Prime Minister has asked me to reply to your telegram of 5 August about the Tongariro power scheme. I assume that your main concern is the effect of the scheme on fishing, and will confine most of my reply to this aspect.

In the very early stages of the investigations it was realised that if the scheme was to go ahead adequate steps would have to be taken to safeguard the world-famous fishing potential of this area. Contact was established then with the Departments concerned with inland fisheries, and has continued ever since. The Marine Department has carried out a tremendous amount of biological research over this period, culminating in the recent release of a most comprehensive technical report 'Fisheries Aspects of the Tongariro Power Development Project.' I understand that the Marine Department has kept fishing interests generally informed of developments through the meetings of the Freshwater Fisheries Advisory Council.

On 3 August a meeting was held in Wellington to discuss the fishing aspects of the Tongariro scheme with representatives of the three Acclimatisation Societies directly concerned, and the Federation of Lake Taupo Fishing Clubs. The steps being taken to safeguard the fishing were described and the following assurances, which I have since confirmed in writing, were given to the societies:

1. Sufficient water will be spilled at the Poutu canal intake to provide the recommended mean flow of approximately 1000 cubic feet per second in the Tongariro River at Tūrangi Bridge.

2. The principal of providing artificial freshes in the Tongariro River to give the best possible fishing conditions is confirmed. The Department will supplement the natural flow of the river with the extra water from the Moawhango catchment and with water released from storage to give as far as the available water will permit the recommended pattern of flow.

3. The technical problems involved in deciding the best position for the Tokaanu Power Station tailrace outlet are still being investigated. If the tailrace discharges into Waihi Bay the Government accepts responsibility for dealing with any consequent problems which may arise in the silting up of the Tongariro River delta.

4. The contaminated Whangaehu water will be entirely excluded from the scheme.

5. In dry spells the flow in the Wanganui and Whakapapa Rivers will not be allowed to fall so low that the safety of the fish is endangered even if this means that the diversion has to be temporarily discontinued

6. Collaboration between the New Zealand Electricity Department and the Departments concerned with fishing will continue into the future and operating procedures will be modified [where] necessary in the light of experience.

The officers of the Marine and Internal Affairs Departments who were present at the [meeting] stated that they were more than satisfied with these safeguards and consider that the fishing potential of the area will not be substantially altered by the power scheme. I would suggest that you might approach the Marine Department for a copy of the fisheries report and obtain their comments at first hand. In the meantime I enclose copies of two press statements recently issued by the Minister of Marine and myself.

I [cannot] agree with your assertion that the power scheme has been suddenly imposed. The proposals have been public knowledge for a number of years and as I have said earlier I understand fishing interests have been kept generally informed by the Marine Department. The scheme received some publicity in 1963 when the Power Planning Committee recommended it for approval and again in March, 1964 when Government approved it in principle subject to being satisfied that suitable arrangements can be made to preserve the interests of parties who would be adversely affected by the scheme. Discussions with these parties have been going on ever since and Government has not yet given its approval for construction to commence.

In reply to your request for alternative schemes to be investigated I am enclosing a copy of the report of the Planning Committee on Electric Power Development in New Zealand. You will see from this that possible alternatives were very fully considered by the committee but were found to have serious economic or technical disadvantages. I think this report will also convince you of the magnitude of the task the Government is facing in providing for the future power needs of the country.

Yours faithfully

T P Shand
Secretary for Marine,
WELLINGTON

Appendix C: Tongariro River pools

Naming of Tongariro River pools

As the fishery developed and anglers returned to fish the same locations year after year, each pool in the river came to be named. The common practice was for some prominent local geographical feature to be used; for instance, the Cherry Pool near the delta was so-called because it had a cherry orchard nearby. The Log Pool obtained its name for obvious reasons, while others were named after angling personalities. The Jones Pool had Robert Jones, local lodge owner and King of Tokaanu, as its namesake. The Major Jones Pool, above the main highway bridge, bears the name of the man who is said to have discovered it, Major Rhys William Wykeham Jones.

Pool maps appear

Although the pool names gained common acceptance among anglers in the early twentieth century, a map was not produced as an official guide for some 20 years. The first list of pools was compiled in 1928 by Eric Mowbray Morilleau, a government surveyor, after title to the Taupō riverbeds was transferred to the Crown. Presumably Morilleau consulted the Department of Tourist and Health Resorts and experienced Tongariro River anglers for the location of each pool or run and its name. The names appeared on official maps published by the New Zealand Lands and Survey Department, the first one appearing in 1929.

The Honorary Geographic Board of New Zealand was formed provisionally in 1924,[1] although it seems to have received no publicity until December 1925.[2] It remained an advisory body for some 20 years, with no statutory authority to make its decisions official. It became the arbiter on New Zealand place names. Because pools on the Tongariro were named well before the board started its work, those names were applied on an informal basis. The board did not publish its first list of confirmed New Zealand place names until 1934, and it was not until 1946 that it received formal recognition under its own Act of Parliament, becoming the New Zealand Geographic Board. However, the only decision it seems to have made affecting the Tongariro river was one in 1948 confirming the places where the Tongariro began and ended.[3]

As the river altered its course with each flood, existing pools and reaches disappeared and were replaced by new ones. Some were an improvement, others were not. With each change came the potential for a new name, which may explain the variations.

From time to time people would produce maps showing fishing pools on the river. While these had no official status, they were useful as a guide. For example, in their book on the Tongariro River[4] Barbara and Allan Cooper refer to a map made in 1932 by a group of anglers from the Upper Waikato and Tongaririo Anglers Club (one of whom included Cecil Whitney). Harold Hickling's *Freshwater Admiral — Fishing the Tongariro River and Lake Taupo*, published in 1960, has a list of pools inside the back cover. In 1959 the Lands and Survey Department issued a map showing new pools that emerged after the 1958 flood (see map C3 on page 298).

There matters rested until Wallace Bain and Barry Greig wrote *Fishing Guide to the Tongariro River and Lake Taupo's Southern Shore* in 1983.[5] The authors, aware that anglers were using outdated information, started with an aerial photograph and then traversed the full length of the Tongariro River, by raft and in waders, in order to accurately plot the position of the pools. After showing the results to anglers for confirmation, they then prepared a map with assistance from a surveyor and submitted it to the New Zealand Geographic Board. The board approved the names (including some new ones on the lower river), which were officially listed in the *Gazette* in 1987. One condition was that an existing name for a pool could not be used if the location of the pool had shifted.

Department of Conservation maps

After the big flood in early 2004, the Department of Conservation released a new map of river pools for anglers. This map, prepared by Colin Lawrence, is based on aerial photographs the department made after working with Environment Waikato and Genesis Energy Ltd.[6] In 2016, the Advocates for the Tongariro River and Ross Baker, proprietor of the Tongariro River Motel, pointed out that 22 pools had ceased to exist. The New Zealand Geographic Board accepted their submission and discontinued the pool names, although 35 others remained and have official status.[7] The New Zealand Geographic Board regards Tongariro River pools as minor features because they are ephemeral. Under board policy their names are recognised if they have sufficient historic or cultural significance.[8] The possessive form of names is to be avoided, and, even if used, an apostrophe is not,[9] examples being Waddells Pool and Admirals Pool.

Another Department of Conservation map issued in March 2018 (see map C5 on page 300) was largely the same as the 2004 one, but with six names omitted: the Big Bend, Cliff, Bypass, de Lautours, Poplar and Cherry.

Whether pool names are official or not they will continue to be adjusted because of the natural dynamics of the river.

C1: Tongariro River pools listed by Eric Morilleau, 1928

(Upper river to lower river)
Sand
Blue
Cliff
Dreadnought
Poutu
Kowhai Flat
Duchess
Sly Grog
Cattle Rustlers
Cicada
Admiral's
Major Jones
Doctor's
Judges
Daisy
Hut
Boat
Duff's
Log
Crescent
Toi
Hingaia
Smallman's
Wanakau
De Lautours
Nihorika
Downs
Bend
Grace's
Te Manuka
Jellicoe
Shingle
Te Piniunui

* Reproduced in Allan and Barbara Cooper, *Pools of the Tongariro: Some History and Humour*. Tūrangi: Turangi District Historical Society, 1975.

C2: Taupo fisheries map, Department of Lands and Survey, NZMS 53 issued in 1929 showing Tongariro River pools

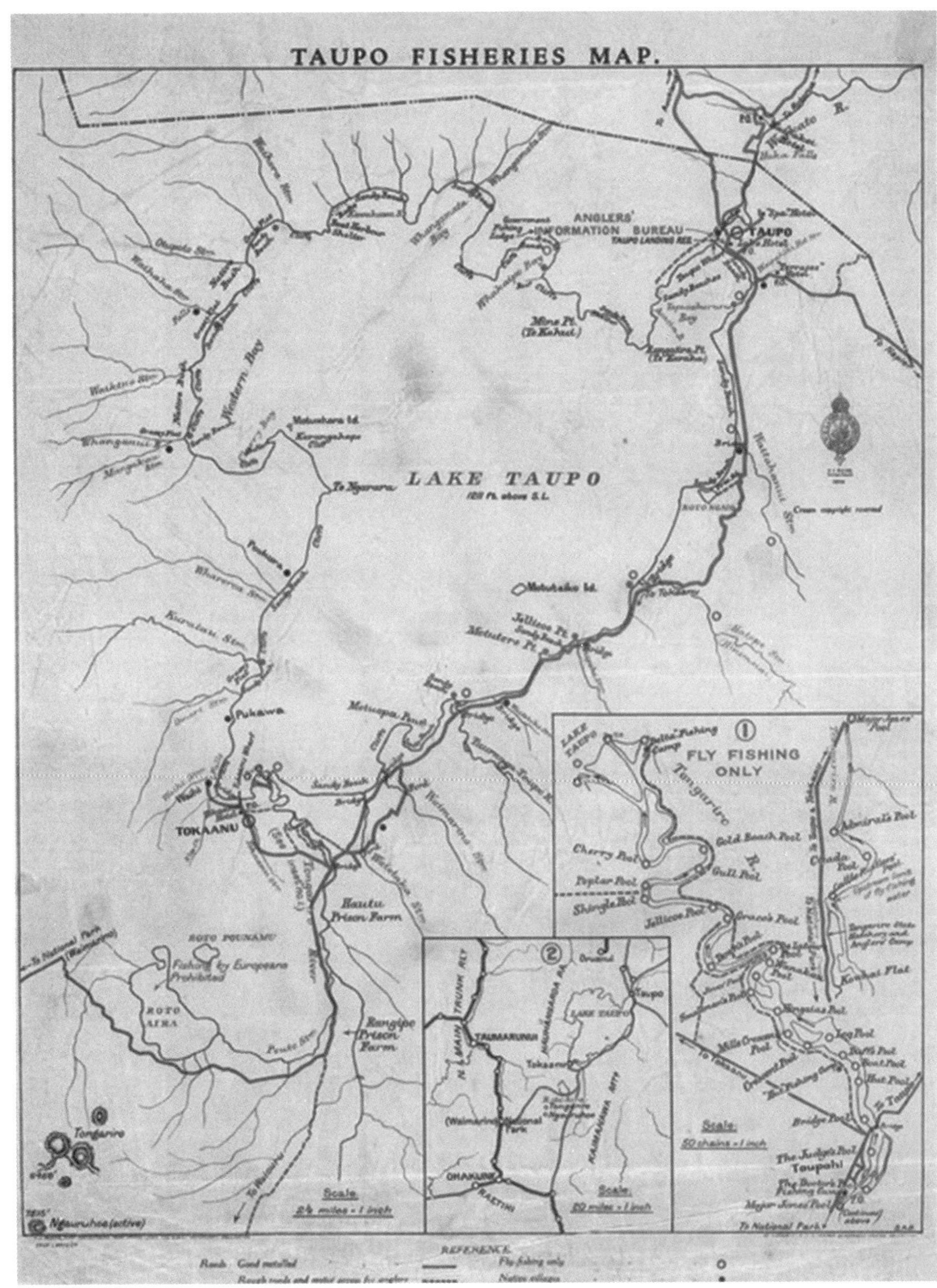

LINZ. Crown Copyright reserved.

C3: Lands and Survey Department map of Tongariro River pools, 1 April 1959

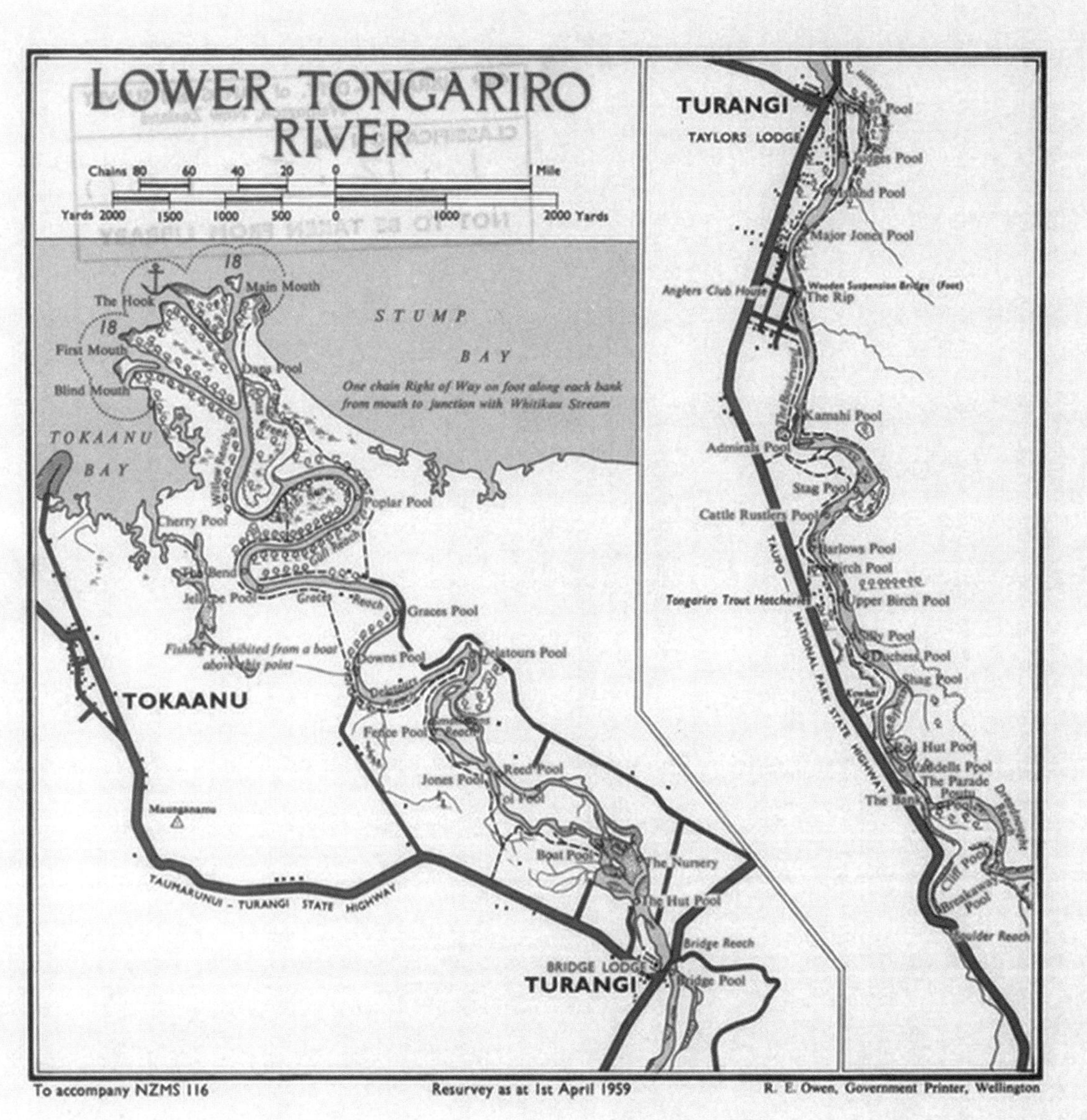

C4: Feature names (fishing pools) on the Tongariro River, as shown on NZMS 284 'Lake Taupo' map, edition I.

1987 pools reordered (upper river to lower)

Beggs Pool
Whitikau Pool
Sand Pool
Blue Pool
Boulder Pool
Big Bend Pool
Fan Pool
Breakaway Pool
Cliff Pool
Poutu Pool
Waddells Pool
The Bypass
Red Hut Pool
Shag Pool
Kowhai Flat
Duchess Pool
Silly Pool
Upper Birch Pool
Lower Birch Pool
Barlows Pool
Cattle/Rustlers Pool
Stag Pool
Admirals Pool
Kamahi Pool
Never Fail Pool
The Boulevard
Hydro Pool
The Rip
Breakfast Pool
Major Jones Pool
Island Pool
Judges Pool
Lonely Pool
Groin Pool
Bridge Pool
Swirl Pool
The Stones
Upper Island Pool
Bain Pool
Shaw Reach
Log Pool
Reed Pool
Jones Pool
The Parade
Fence Pool *
Smallmans Reach
The Bends Pool
De Latours Pool
De Latours Reach
Downs Pool
Graces Pool
Graces Reach
Jellicoe Pool
The Bend
Gull Reach
Poplar Pool
Poplar Run
Cobham Pool
Cherry Pool
Willow Reach
Dans Pool

Pool names officially discontinued by New Zealand Geographic Board, 30 June 2016

Bridge Pool
Cherry Pool
Cobham Pool
Dans Pool
De Latours Pool
De Latours Reach
Graces Pool
Graces Reach
Groin Pool
Gull Reach
Jellicoe Pool
Jones Pool
Poplar Pool
Poplar Run
Shaw Reach
Swirl Pool
The Bend
The Bends Pool
The Bypass
The Stones
Upper Island Pool
Willow Reach

Source: *New Zealand Gazette*, No. 163, 24 September 1987, p 4458.

C5: Map of Tongariro River pools, Department of Conservation, March 2018

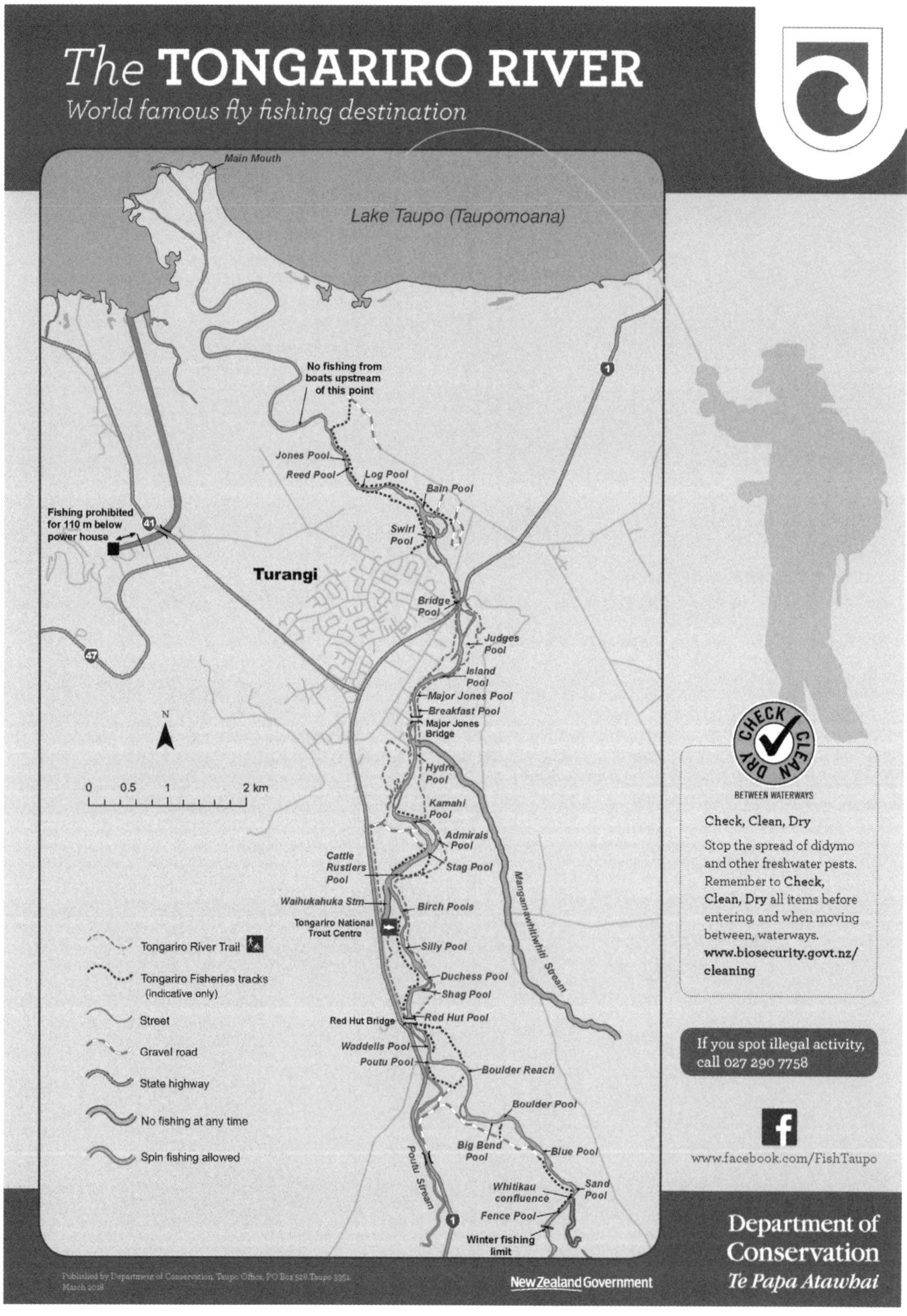

Department of Conservation.

Acknowledgements

Many people gave assistance and encouragement to me in the writing of this book. Grateful thanks are due to the late John Parsons (Taupō) for providing photographs and useful references, Professor Paul Williams (University of Auckland) who advised on the geology of the Tongariro region and freshwater scientist Michel Dedual (Tūrangi).

Others who helped were Valentine Were (Takapuna), Helen Lewis (Waikanae) for information on E. J. Wiffin, Norrie Ewing (Hokitika) for information on Pat Burstall, Margaret Maynard (Plimmerton), Mavis Hall (Auckland), Alpha Cuthbertson, former proprietor of Taylor's Lodge (Tūrangi), Betty Hura-O'Connor (Tūrangi library), Rosemary Hanes (Reference librarian, Library of Congress, Washington DC, USA), Peter Barker (Rotorua), Eugene Fleming (California Department of Fish and Game), Mary Nisbet (secretary/manager of TALTAC), John Gibbs (Department of Conservation Officer, Tongariro/Taupō Conservancy, Tūrangi), Mary Ellen Wilson (New Zealand Services librarian, Rotorua Public Library), Hono Lord (Tūrangi), the House of Hardy Ltd (London), Alan Hayton (New Plymouth), Grey Whitney (Auckland), Del Little (Taupo District Library), Dr A. R. Morton (Archivist/Deputy Curator, Royal Military Academy, Sandhurst, England), Diana Porter (Secretary, Bath Photographic Society, England), Rosemary Tiley (Bath, England), Ian Walker (Penrith, Cumbria, England), Paul Arengo-Jones (London, England), David Holdsworth (Yorkshire, England), Nicola Stewart (Taupo Museum and Art Gallery), the family of the late Val Raymond and Philippa Tucker (Featherston).

I researched records held by several institutions and organisations. Valuable guidance came from staff at the National Library (Wellington) and Archives New Zealand (Wellington and Auckland), the New Zealand Film Archive, the Taupō District Library, the Tongariro National Trout Centre and committee members of the Advocates for the Tongariro River (Inc).

Special mention should be made of the National Library website Papers Past, which is a boon to researchers.

To get the book to its final form took the dedicated efforts of Louise Russell and Adrian Kinnaird of David Bateman Ltd, along with subeditor Kate Stone and indexer Louise Raynes. For any errors in the text I must take responsibility.

Finally, my sincere thanks to my family for their kindness and support during the long years of the writing project.

Grant Henderson
Northcote
Auckland

Endnotes

Abbreviations

AJHR	*Appendix to the Journals of the House of Representatives*
AS	*The Auckland Star*
BPT	*The Bay of Plenty Times*
Dom	*The Dominion*
EP	*The Evening Post*
HBH	*The Hawke's Bay Herald*
JRSNZ	*Journal of the Royal Society of New Zealand*
MS	*The Manawatu Standard*
MT	*The Manawatu Times*
NZFSG	*New Zealand Fishing and Shooting Gazette*
NZG	*New Zealand Gazette*
NZH	*The New Zealand Herald*
NZM	*The New Zealand Mail*
NZO	*New Zealand Outdoor*
NZPD	*New Zealand Parliamentary Debates*
NZT	*The New Zealand Times*
ODT	*The Otago Daily Times*
OW	*Otago Witness*
PBH	*The Poverty Bay Herald*
RDP	*Rotorua Daily Post*
TT	*Target Taupō*
WC	*Wanganui Chronicle*
WDT	*Wairarapa Daily Times*
WH	*The Wanganui Herald*

Chapter 1: The geography of the Tongariro River

1 Professor Paul Williams, PhD, ScD (Cambridge), School of Environment, University of Auckland, personal communication with author.
2 *NZG*, 29 July 1948, p 962.
3 J. C. Andersen also notes that early surveyors' maps (1876 and 1912) gave the name Tongariro to different streams from the Kaimanawa Range; see *The Evening Post*, 14 November 1931 at p 12.
4 Ferdinand von Hochstetter, *New Zealand: Its Physical Geography, Geology and Natural History*, Stuttgart: J. G. Cotta, 1867, at p 371.
5 Ferdinand von Hochstetter, *New Zealand: Its Physical Geography, Geology and Natural History*, Stuttgart: J. G. Cotta, 1867, at p 371.
6 P. Froggatt, Volcanic Hazards at Taupo Volcanic Centre. Volcanic Hazards Information Series No. 7. Palmerston North: Ministry of Civil Defence, 1997.
7 Opus International Consultants Limited, Taupo District Flood Hazard Study: Tongariro River. Wellington: Opus International Consultants Limited, July 2011, at p 3.
8 V. E. Neall, B. F. Houghton, S. J. Cronin and A. R. Mitchell, Volcanic Hazards at Ruapehu Volcano. Volcanic Hazards Information Series No. 8. Wellington: Ministry of Civil Defence, 1999.
9 Paul W. Williams, New Zealand Landscape: Behind the Scene. Oxford: Elsevier, 2017, p 225.
10 M. Dedual and E. Cudby, 'Potential effects of volcanic events on the Taupo trout fishery', *TT*, 30, March 1999, at p 4.

Chapter 2: Establishment of the fishery

1 The confiscations were authorised by punitive legislation in the form of the New Zealand Settlements Act 1863 and the Suppression of Rebellion Act 1863: 'The Explosive Frontier'. In: B. Dalley and G. McLean (eds), *Frontier of Dreams: The Story of New Zealand* (pp 127-153). Auckland: Hodder Moa, 2006, at p 144.
2 *AS*, 12 June 1884, p 2.
3 *HBH*, 9 April 1883, p 2.
4 'The Explosive Frontier', at p 152.
5 J. Belich, *Making Peoples: A History of the New Zealanders from Polynesian Settlement to the End of the Nineteenth Century*. Honolulu: University of Hawai'i Press, 1996, p 258
6 Matthew Wright, *Illustrated History of New Zealand*. Auckland: David Bateman Ltd, 2013, at p 199.
7 Noted in an address by A. E. Hefford, *EP*, 7 April 1933, at p 9. Hefford states the browns were first established by A. M. Johnson of Canterbury. An earlier attempt in 1864 using ova from the United Kingdom was unsuccessful.
8 The first brown trout ova brought to New Zealand arrived in September 1867 in the care of Alec Johnson, curator of the North Canterbury Acclimatisation Society; ODT, 20 September 1867, p 5.
9 *AS*, 13 August 1898, p 1.
10 J. Te H. Grace, *Tuwharetoa: A History of the Maori People of the Taupo District*. Wellington: Reed Books, 1959, p 515.
11 D. Rowe and I. Kusabs, *Taonga and Mahinga Kai of the Te Arawa lakes: A Review of Current Knowledge — Koaro*, NIWA Client Report HAM2007-022. Hamilton, National Institute of Water and Atmospheric Research, 2007.
12 *TT*, March 2003, Issue 42, p 30.
13 *BPT*, 20 January 1880, p 2.
14 *NZH*, 28 March 1881, p 3.
15 *ODT*, 3 August 1872, p 6.
16 *ODT*, 12 March 1890, p 2.
17 Temple Sutherland, in *Maui and Me*. Wellington: A. H. & A. W. Reed, 1963, p 122), notes the carp liberations.
18 *NZH*, 6 July 1887, p 5.
19 *NZH*, 2 September 1885, p 5; 11 March 1886, p 6.
20 P. Burstall, 'Trout fishery — history and management'. In: D. J. Forsyth and C. Howard-Williams (co-ordinators), *Lake Taupo: Ecology of a New Zealand Lake* (pp 119–131). DSIR Information Series No. 158.Wellington: Science Information Publishing Centre, DSIR, 1983, at p 120.
21 John Parsons, *A Taupo Season*. Auckland: Collins, 1979, at p 13.
22 *The Daily Telegraph* (Napier), 29 March 1894, p 2.
23 *NZH*, 13 December 1892, p 6.
24 *NZH*, 29 January 1901, p 6.
25 *HBH*, 14 June 1893, p 3.
26 Annual report of the Wellington Acclimatisation Society for 1893, noted in the *Otago Witness*, 4 May 1893, p 33.
27 *EP*, 12 August 1895, p 3. Presumably these were brown trout eggs.
28 *WDT*, 12 August 1895, p 2.
29 *Hot Lakes Chronicle*, 23 January 1897, p 2.
30 *OW*, 6 February 1896, p 37.
31 *AS*, 12 March 1896, p 5.
32 *The Timaru Herald*, 20 March 1896, p 2.
33 *EP*, 24 March 1900, p 2.
34 Evidence of Thomas Ryan to Public Petitions A to L

Committee, 24 July 1913, *AJHR*, I-IA, p 10.
35 *Hawera and Normanby Star*, 2 July 1901, p 2.
36 *Taupo and Turangi Weekender*, 23 March 2017, p 9.
37 Charles Bruce Morison (later Morison KC) was a Wellington lawyer and authority on company law.
38 *EP*, 9 April 1904, p 14.
39 *WH*, 12 January 1906, p 4.
40 *EP*, 23 February 1907, p 14.
41 *EP*, 26 January 1907, p 14.
42 *EP*, 6 March 1909, p 14.
43 *NZH*, 14 June 1883, p 5.
44 See Bob McDowall, *Essays of a Fishery Scientist: 50 Years of Experience*, compiled and edited by D. Jellyman. NIWA Information Series No. 80. Wellington: NIWA, 2011, p 12.
45 D. Scott, J. Hewitson and J. C. Fraser, 'The origins of rainbow trout, *Salmo gairdneri* Richardson, in New Zealand', *California Fish and Game*, 1978, 64(3): 210–218. In the 1980s the American Fisheries Society decreed the name of the species should be changed to *Oncorhynchus mykiss*, the former word being the genus attributed to salmon and trout.
46 Druett, *Exotic Intruders*, p 97.
47 *EP*, 27 September 1935, p 16.
48 *HBH*, 13 June 1896, p 4.
49 *WDT*, 14 March 1898, p 3.
50 *TT*, March 1998, Issue 27, p 4.
51 *WC*, 15 March 1898, p 2; *Target Taupō*, March 1998, Issue 27, p 5.
52 B. Cooper, *The Remotest Interior: A History of Taupo. Tauranga:* Moana Press, 1989, p 112.
53 W. A. Sullivan, *Changing the Face of Eden, A History of the Auckland Acclimatisation Society 1861–1990*. Hamilton: Auckland/Waikato Fish and Game Council, 1998, p 84.
54 The Hawke's Bay Acclimatisation society annual report for 1902/03 records that 10,000 rainbow fry were liberated in the Ōtamatea River, a Rangitaiki tributary located between Napier and Taupō (*HBH*, 29 May 1903, p 4).
55 *The Hawke's Bay Herald*, 26 June 1901, p 4.
56 *The Poverty Bay Herald*, 4 May 1901, p 4. Mr Grant of Whirinaki Station reportedly stocked Whirinaki River with 350 rainbow trout in 1897. By 1900 he was catching double-figure fish.
57 *NZH*, 12 June 1911, p 9.
58 *EP*, 27 September 1935, p 16; C. R. Ashby, *The Centenary History of the Auckland Acclimatisation Society 1867–1967*. Auckland: Auckland Acclimatisation Society, 1967, pp 116–117; *TT*, December 2002, Issue 41, p 6.
59 *AS*, 2 August 1905, p 9; *The Waikato Independent*, 11 July 1911, p 5; *NZH*, 2 October 1909, p 8.
60 *TT*, March 1998, Issue 27, p 5.
61 *NZM*, 10 December 1902, p 67; *HBH*, 1 May 1903, p 2.
62 *Yorkshire Post*, 7 September 1929, p 16.
63 Druett, *Exotic Intruders*, p 138.
64 *Observer*, 24 March 1906, p 3. In 1998 the New Zealand Court of Appeal ruled that Māori fishing rights (preserved under the Conservation Act 1987) did not extend to catching trout: see *McRitchie v Taranaki Fish and Game Council* [1998] NZCA 203; [1999] 2 NZLR 139 (CA).
65 *NZH*, 29 March 1923, p 6.
66 *NZT*, 31 July 1911, p 3.
67 Cooper, *The Remotest Interior*, p 113.
68 *AS*, 11 August 1934, p 12.
69 *WDT*, 1 February 1905, p 3.
70 *EP*, 23 February 1907, p 14.
71 *The Bush Advocate*, 30 November 1907, p 4.

Chapter 3: Catching trout by the ton

1 *MT*, 14 February 1903, p 3.
2 Glenn Maclean, 'Rainbow trout in Taupō waters', *TT*, 27 March 1998, Issue 27, p 5; *AS*, 4 August 1911, p 9.
3 John Pascoe, 'Historical Record — Thermal Fisheries', Alexander Turnbull Library MS Papers -6523.
4 *EP*, 21 March 1903, p 14.
5 *NZH*, 17 September 1906, p 4.
6 *The Bush Advocate*, 26 April 1907, p 2.
7 *WH*, 30 September 1907, p 4.
8 *OW*, 29 January 1908, p 36.
9 *EP*, 23 February 1907, p 14.
10 *EP*, 19 January 1907, p 14.
11 *NZH*, 29 April 1908, p 8.
12 *NZH*, 6 January 1909, p 7; 12 January 1909, p 6.
13 *NZH*, 15 March 1911, p 10.
14 *PBH*, 1 April 1912, p 5.
15 C. E. Lucas, 'Trout fishing in New Zealand', in Sir Herbert Maxwell (ed), *Fishing at Home and Abroad* (pp 138–143). London: The London and Counties Press Association Ltd, 1913.
16 *EP*, 12 May 1909, p 4.
17 *WC*, 22 January 1909, p 4.
18 *NZH*, 17 February 1914, p 8.
19 *NZH*, 24 December 1913, p 10.
20 *HBH*, 29 September 1880, p 3.
21 *NZH*, 12 May 1921, p 9.
22 *EP*, 11 July 1910, p 3.

Chapter 4: The first Tongariro anglers

1 *Hawera and Normanby Star*, 17 February 1905, p 2.
2 *OW*, 15 February 1905, p 59.
3 *The Referee* (NSW), 12 June 1912, p 11.
4 John Parsons, *The Fishing Years*. Taupō: Acacia Bay Books, 2004, p 179.
5 *The Western Times* (England), 8 May 1906, p 5.
6 A. G. C., 'With rod and bicycle in Norway', *The Fishing Gazette*, 1908, p 172 (https://archive.org › stream › fishinggazettede5619unsc_djvu)
7 *MT*, 20 April 1909, p 4.
8 *NZM*, 24 April 1907, p 11.
9 Ibid.
10 Ibid.
11 *EP*, 23 March 1907, p 14.
12 *The Field*, quoted in *The Waikato Argus*, 31 July 1907, p 2.
13 *EP*, 23 March 1907, p 14.
14 *EP*, 13 April 1907, p 14.
15 *NZFSG*, 1 February 1929, p 8.
16 *Dom*, 17 June 1910, p 9.
17 *NZH*, 14 June 1911, p 4.
18 Shilson's fellow anglers included J. C. Buckingham and Arthur M. Naylor (*Wanganui Chronicle*, 27 March 1915, p 4).
19 *NZH*, 14 June 1911, p 4.
20 *NZFSG*, December 1927, p 3.
21 *NZH*, 11 April 1914, p 8.
22 Cooper, *The Remotest Interior*, p 114; *Otago Witness*, 9 March 1910, p 64.
23 Harold Hickling, *Freshwater Admiral — Fishing the Tongariro River and Lake Taupo*. Wellington: A. H. & A. W. Reed, 1960, p 122.
24 *TT*, January 2012, Issue 64, p 46.
25 1871 census of Bath, England.
26 British Army lists, The National Archives, London. His will,

signed in 1884, shows his rank as Lieutenant.
27 General Gordon was killed by Moslem forces who had revolted and taken control of the Sudan. After Khartoum was retaken by British forces, it emerged that, had the relief column arrived two days earlier than it did, Gordon would very likely have been saved. Alan Moorehead gives an excellent account of the matter in *The White Nile* (Harmondsworth: Penguin Books, 1963).
28 *The Times*, 24 June 1885, at p 5 refers to Lt R. W. W. Jones, 1st battalion, Royal Sussex Regiment.
29 *The Times*, 7 December 1892, p 7.
30 *The Times*, 21 September 1901, p 6.
31 *The Times*, 10 June 1905, p 10; *Target Taupō*, January 2012, Issue 64, p 46.
32 Bath and County Club records. Shipping passenger lists show he travelled from New Zealand to London in May 1908 (see Ancestry.com).
33 *NZT*, 14 March 1907, p 5.
34 *New Zealand Free Lance* 28 December 1907, p 4.
35 *AS*, 13 May 1907, p 2.
36 *EP*, 5 March 1908, p 6.
37 *AS*, 13 May 1907, p 2.
38 *NZH*, 7 April 1909, p 8.
39 *EP*, 13 October 1911, p 7.
40 *AS*, 21 March 1912, p 5. See also 18 December 1912, p 7.
41 *OW*, 9 March 1910, p 64.
42 Cooper, *The Remotest Interior*, p 114.
43 *MT*, 8 November 1924, p 11.
44 *AS*, 10 April 1912, p 4. (J. C. Buckingham was with him.)
45 *BPT*, 26 April 1921, p 2.
46 *The Times*, 29 September 1922, p 13, under 'Wills and Bequests' notes the will of Major Rhys Wykeham Jones, Royal Sussex Regiment (Ret'd) of Mount Royal, Bournemouth. Gross value of his estate was £14,193.
47 *WC*, 22 January 1909, p 4.
48 *The Field, the Country Gentleman's Newspaper*, Vol 114, 14 August 1909, p 302.
49 *The Waikato Argus*, 16 May 1912, p 4.
50 *MS*, 28 October 1912, p 5.
51 *NZFSG*, 2 January 1928, p 12.
52 *Who's Who*. London: A. & C. Black, 1920.
53 *Australasian*, 23 November 1918, p 18.
54 *EP*, 29 November 1916, p 8; *The New Zealand Herald*, 2 April 1917, p 6.
55 *Australasian*, 23 November 1918, p 18.
56 *NZFSG*, 2 January 1928, p 12.
57 Ibid.
58 *AS*, 26 March 1928, p 8.
59 *The Feilding Star*, 8 November 1929, p 4.
60 *NZH*, 22 November 1933, p 10.
61 Ibid, p 12.
62 *AS*, 21 November 1933, p 7.
63 *AS*, 20 October 1933, p 3.
64 *NZH*, 22 November 1933, p 6. He left an estate worth £23,056.
65 *Northern Advocate*, 30 July 1945, p 3.
66 *Northern Advocate*, 27 March 1911, p 4.
67 *Dom*, 24 May 1911, p 6.
68 *NZH*, 15 January 1906, p 5.
69 *NZFSG*, December 1927, p 15.
70 *The Press*,7 January 1913, p 7.
71 *Rainbow Trout Centenary: Lake Taupo New Zealand 1898–1998*. Taupo: Harland-Baker Publishing Ltd, 1998, p 20. Mrs Potts also features in the supplement to the *Auckland Weekly News*, 18 May 1911, p 8; 28 December 1911, p 9; *OW*, 6 November 1912, p 47.

Chapter 5: Robert Jones's private fishery

1 *BPT*, 30 October 1905, p 5, contains a liquor licensing notice stating Robert Jones was the owner of the Tokaanu Hotel premises.
2 Allan and Barbara Cooper, *Pools of the Tongariro: Some History and Humour*. Tūrangi: Turangi District Historical Society, 1975, p 3.
3 *Dom*, 5 January 1910, p 4.
4 *NZH*, 4 May 1905, p 8.
5 *OW*, 3 November 1909, p 64.
6 *NZFSG*, 2 January 1928, p 12.
7 *NZFSG*, December 1927, p 2.
8 *NZFSG*, December 1927, pp 2–3.
9 *Dom*, 20 February 1911, p 6.
10 Ibid.
11 *NZH*, 23 March 1932, p 2.
12 C. H. M. Thring, *The Trials and Pleasures of an Uncompleted Tour*. London: Simpkin, Marshall, Hamilton, Kent & Co, 1914.
13 *NZH*, 17 September 1906, p 4.
14 *MS*, 15 January 1908, p 4.
15 The charge was 2 shillings 6 pence per day, or 10 shillings per year, according to Cecil Whitney; see *EP*, 2 December 1902, p 12.
16 The Native Land Act 1909 allowed the Crown to issue proclamations preventing private interests from bidding for native land the Crown wished to acquire. From 1918, proclamations were issued over much land in the Taupō district, restricting alienation. See 'Ngāti Tūwharetoa and Te Kota hitanga o Ngāti Tūwharetoa and the Crown, Deed of Settlement, 8 July 2017', para 2.254.
17 *NZH*, 24 September 1924, p 10.
18 F. Carr Rollett, *Angling in New Zealand*. Auckland: Whitcombe and Tombs Ltd, 1923, p 61.
19 *Dom*, 20 March 1909, p 4.
20 *NZH*, 5 September 1923, p 8.
21 Allan and Barbara Cooper, *Pools of the Tongariro*, p 29.

Chapter 6: Decline and recovery

1 *PBH*, 31 May 1911, p 8.
2 *EP*, 17 June 1913, p 3.
3 *The Press*, 2 February 1907, p 3.
4 '7th annual report of Tourist and Health Resorts Department', *AJHR*, 1908, H-2, p 30.
5 *AJHR*, 1909, H-2, p 8.
6 R. R. Strickland, 'Pre-European transfer of smelt in the Rotorua-Taupo area, New Zealand', *JRSNZ*, 1993, 23(1), 13–28. DOI: 10.1080/03036758.1993.10721214.
7 *The Dominion*, 14 May 1912, p 9.
8 *The Dominion*, 17 May 1912, p 5.
9 By 1912 the role of the shag as original host for the parasitic nematode worm had been confirmed; see *AJHR*, 1913, H-15B, p 19. In 1917 confirmation also came from the report by Dr Reakes, 'The worm parasite in trout', 3 July 1917, AJHR, 1917, H-22.
10 *TT*, March 1995, Issue 18, p 6.
11 *NZH*, 16 April 1912, p 5.
12 *PBH*, 22 May 1912, p 7.
13 *WH*, 30 January 1909, p 4.
14 'Report of L. F. Ayson to Marine Department', *AJHR*, 1913, H-15B, p 19.
15 *ODT*, 22 May 1912, p 8.
16 'Report of L. F. Ayson to Marine Department', *AJHR*, 1913, H-15B, p 18.
17 'Evidence of F. Moorhouse given to Parliamentary committee, 24 July 1913', *AJHR*, 1913, I-1A, p 16.

18 G. Stokell, 'The nematode parasites of Lake Ellesmere trout', *Transactions and Proceedings of the Royal Society of New Zealand*, 1937, p 66, 80–96.
19 *NZT*, 10 October 1919, p 6.
20 'Preliminary report of Professor E. E. Prince to the New Zealand government, 14 August 1914', *AJHR*, 1914, H-15c, p 20.
21 *EP*, 25 July 1913, p 3.
22 Report of the Conservator of Fish and Game, *AJHR*, 1917, H-22, p 23.
23 Figures in the table are from reports the Internal Affairs Department made each year to Parliament; see for example 'Report on Fisheries at Lakes Taupo and Rotorua from 1st June 1914', *AJHR*, 1915 Session I, H-21.
24 *EP*, 13 April 1914, p 7.
25 *WDT*, 1 April 1915, p 4.
26 *The Star*, 31 March 1916, p 5.
27 'Annual report of Dept of Internal Affairs', *AJHR*, 1922, H-22, p 9.
28 *The Mid-Sussex Times*, 30 December 1919, p 2.
29 *The Staffordshire Advertiser*, 16 October 1920, p 5.
30 *NZH*, 25 March 1922, p 8.
31 'Report of the Conservator of Fish and Game, 22 July 1921', *AJHR*, 1921 Session I-II, H-22, p 28.
32 Ibid.
33 'Annual report of Department of Tourist and Health Resorts', 1923 Session I-II, *AJHR*, H-02, p 3.
34 *AS*, 5 May 1923, p 9.
35 *Manawatu Evening Standard*, 10 November 1923 , p 5.
36 'Annual report of Department of Tourist and Health Resorts', 1924 Session I, AJHR, 1924, H-02, p 3.
37 *Australasian*, 1 December 1923, p 27.
38 Ibid.
39 *NZH*, 16 May 1924, p 8.
40 *NZH*, 15 January 1924, p 6.
41 *MT*, 18 July 1924, p 2.
42 John Pascoe, *History of Thermal Fisheries*, Alexander Turnbull Library MS-papers 6523, p 44.
43 Burstall, 'Trout fishery — history and management', P 124.
44 *NZH*, 18 March 1924, p 8.
45 *NZH*, 21 April 1927, p 8.
46 *NZFSG*, 2 April 1928, p 20.
47 *The Blue Mountain Echo*, 10 June 1927, p 2.
48 *NZH*, 25 May 1929, p 12.
49 *NZO*, September 1949, p 7.
50 P. Burstall, *Taupo Fishery Report*, p 8; Archives NZ, IA, 1, W2578, 47/8.
51 Letter from J. S. W. Neilson to Andy Kean, 12 June 1940, Archives NZ, IAD 79/63.
52 Department of Internal Affairs, annual report for year ended 31 March 1937, *AJHR*, 1937 Session I, H-22, p 11.

Chapter 7: Zane Grey in New Zealand

1 Steve Netherby, 'Zane Grey: author, angler, explorer', *Field and Stream*, January 1972, p 52.
2 See Carole van Grondelle, *Angel of the Anzacs: The Life of Nola Luxford*. Wellington: Victoria University Press, 2000.
3 Stephen J. May, *Maverick Heart: The Further Adventures of Zane Grey*. Athens, OH: Ohio University Press, 2000, p 174; Zane Grey, *Tales of the Angler's El Dorado: New Zealand*. London: Hodder & Stoughton, 1926, p 215 and the author's dedication.
4 *AS*, 28 August 1925, p 5.
5 May, *Maverick Heart*, p 174.
6 Department of Tourist and Health Resorts, Archives NZ, TO 1, 40/33.
7 *NZH*, 20 January 1926, p 11; *AS*, 19 January 1926, p 8.
8 *NZH*, 9 April 1926, p 8.
9 Letter by Dolly Grey, 22 March 1927 (quoted in Candace C. Kant (ed.), *Dolly and Zane Grey: Letters from a Marriage*, Reno: University of Nevada Press, 2011, p 257).
10 *The Thames Star*, 8 August 1931, p 3.
11 May, *Maverick Heart*, p 158.
12 *NZH*, 16 April 1926, p 12.
13 *NZH*, 20 April 1927, p 8.
14 A. R. Mills, *Lord Suffer Me to Catch a Fish*. Wellington: A. H. & A. W. Reed, 1967, p 128.
15 Grey, *Tales of the Angler's El Dorado*, p 180.
16 Ibid, p 178.
17 Te Hoka o te rangi Downs (1885–1940), Ngāti Tūwharetoa, was commonly known by the abbreviated name Hoka Downs.
18 Grey, *Tales of the Angler's El Dorado*, p 195.
19 The Maori Land Amendment and Maori Land Claims Adjustment Act 1926.
20 *NZH*, 16 February 1926, p 11.
21 Ibid.
22 George Reiger, *Profiles in Saltwater Angling*. New Jersey: Prentice-Hall Inc, 1973, p 133.
23 *AS*, 26 April 1926, p 10.
24 *NZH*, 14 April 1926, p 15.
25 *NZH*, 19 March 1927, p 13.
26 *NZH*, 5 March 1932, p 12.
27 Refer to the website www.historicfilms.com for the 20-minute film involving Grey's adventures in Tahiti and New Zealand.
28 Grey's second Tongariro expedition spanned the period approximately from 6 April to 18 May, according to his notes in *Tales from a Fisherman's Log* (London: Hodder and Stoughton Ltd, 1978, p 116). It was preceded by two and a half months' big game fishing in the Bay of Islands (p 57). The Tongariro portion of his visit included the Easter holiday in mid-April 1927.
29 *NZH*, 14 March 1932, p 10; Loren Grey, 'Zane Grey's New Zealand', *Fly, Rod and Reel*, November/December 1991, p 56.
30 *NZH*, 21 April 1927, p 8.
31 *AS*, 26 April 1926, p 10.
32 *AS*, 20 January 1927, p 9.
33 *NZH*, 21 January 1927, p 12.
34 *The New Zealand Truth*, 5 May 1927, p 1.
35 *NZH*, 5 January 1927, p 10.
36 *EP*, 22 April 1927, p 8. This report records Romer Grey's catch of a 15¼lb rainbow on a 5½oz rod, from the Dreadnought Pool. The party had caught 80 trout by that date.
37 *NZH*, 25 May 1927, p 12.
38 Regulations for Trout Fishing, Taupo District, r 9(1)(d), *NZG*, 7 October 1926, p 2896.
39 Grey, *Tales from a Fisherman's Log*, p 72.
40 Ibid, p 74.
41 *NZH*, 9 April 1927, p 8.
42 Ibid.
43 Ibid, p 8.
44 Bryn Hamond, *The New Zealand Encyclopaedia of Fly Fishing*. Auckland: The Halcyon Press, 1988, p 77.
45 Matthew Wright, *Illustrated History of New Zealand* (2nd edn). Auckland: David Bateman, 2013, p 313.
46 *NZPD*, 3 September 1926, p 285.
47 Kant, *Dolly and Zane Grey: Letters from a Marriage*, p 247.
48 Grey arrived in New Zealand in mid-December 1928: *The New Zealand Herald*, 18 December 1928, p 12. His first day's big game fishing was in early January 1929. The party left for Tokaanu in mid-March 1929: *NZH*, 14 March 1929, p 10. Just

before departing for Tokaanu, Grey criticised the heavy tackle used by New Zealand anglers for big game fish.
49 Kant, *Dolly and Zane Grey, Letters from a Marriage*, p 313.
50 *EP*, 20 August 1931, p 15.
51 *NZH*, 15 February 1933, p 12.
52 *NZH*, 24 August 1936, p 11.
53 *Field and Stream*, January 1972, p 88.
54 *Fly, Rod and Reel*, November/December 1991, p 58.
55 See Ernest Hemingway, *Green Hills of Africa*. Harmondsworth: Penguin Books, 1966 [1935].
56 *NZH*, 14 March 1932, p 10.
57 Ibid.

Chapter 8: Royal visitors
1 *The Sydney Morning Herald*, 2 March 1927, p 15.
2 *AS*, 2 March 1927, p 10.
3 Hickling, *Freshwater Admiral*, p 126.

Chapter 9: A public fishery emerges
1 Order in Council, 21 September 1886.
2 AS, 29 October 1906, p 4.
3 Regulations for Trout-fishing, Rotorua Acclimatisation District, *NZG*, 29 October 1914, p 3876.
4 R. Galbreath, *Working for Wildlife: A History of the New Zealand Wildlife Service*. Wellington: Bridget Williams Books, 1993, p 8.
5 Ibid.
6 Taupo Trout Fishing Regulations 1939 (No. 200 of 1939), r 1(3).
7 Regulations for Trout Fishing, Rotorua Acclimatisation District 2014, r 2 (Order in Council 27 October 1914).
8 Ibid, r 4.
9 Ibid, r 6. Lake Taupo tributaries were included in the group reserved for fly fishing only until the restriction was removed by Order in Council of 21 December 1914 (*Gazette*, 7 January 1915, p 12).
10 Regulations for Trout Fishing, Rotorua Acclimatisation District 2014, r 5.
11 A. R. Mills recalls doing just that on his first trip to Taupo as a lad; see *Lord Suffer Me to Catch a Fish*, p 17.
12 *NZH*, 21 August 1924, p 10.
13 Native Land Amendment and Native Land Claims Adjustment Act 1924, s 29(3).
14 'Letter to minister, 21 April 1925', Archives NZ, IA 1, 79/19.
15 Michael Bassett, *Coates of Kaipara*. Auckland: Auckland University Press, 1995, p 26.
16 *NZH*, 22 April 1926, p 10.
17 Ibid.
18 'File memo 26 November 1927', Archives NZ, IA 1, 79/31.
19 *ODT*, 27 April 1926, p 3.
20 *NZPD*, 3 September 1926, p 285.
21 *NZH*, 22 April 1926, p 10.
22 *ODT*,6 July 1926, p 13.
23 Ibid.
24 'Memorandum of 9 September 1926 to Conservator of Fish and Game', Archives NZ, MA 31, 23a.
25 See sections 14–16 of the Native Land Amendment and Native Land Claims Adjustment Act 1926 concerning Taupō waters.
26 'Proclamation Declaring Beds of certain Rivers or Streams flowing into Lake Taupo to be Crown Land, and reserving a Right of Way over Land on each Bank of such Rivers or Streams, and restricting the Use of certain Parts thereof'; supplement to the *NZG*, 8 October 1926, p 2895.
27 See proclamation at n 26, Pt IV and Third Schedule.
28 Taupo Trout Fishing Regulations 1926, r 4(4). By r 4(5), it was decreed that a fishing licence gave no access rights apart from those conferred by the regulations unless the landowner consented.
29 *AS*, 11 June 1927, p 12.
30 Taupo Trout Fishing Regulations 1927, supplement to the *NZG*, 29 September 1927, p 3031.
31 Ibid, r 15.
32 Taupo Trout Fishing Regulations 1926, r 3(3).
33 *NZH*, 9 April 1927, p 14.
34 *EP*, 9 April 1927, p 10.
35 *EP*, 17 May 1926, p 8.
36 *EP*, 17 May 1927, p 10.
37 *NZH*, 18 May 1931, p 10.
38 *AS*, 10 May 1927, p 5.
39 Taupo Trout Fishing Regulations 1928, r 7(5).
40 *NZH*, 2 November 1931, p 10.
41 *PBH*, 19 March 1929, p 2.
42 Native Land Amendment and Native Land Claims Adjustment Act 1926, s 14.
43 'Letter from Solicitor-General, 22 December 1928', Archives NZ, IA 1, 79/31.
44 Ibid.
45 Cited in memorandum *Taupo Waters – Fishing Rights*, p 3, Archives New Zealand, IA 1, 79/19.
46 'Letter from Solicitor-General, 16 October 1928', Archives NZ, IA 1, 79/31.
47 'Letter from H. H. Ostler to J. G. Coates, 21 July 1926'; Archives NZ, ACIH 16064, MA31, 13/23a, IA 26/18/6, R22041865.
48 *PBH*, 9 December 1926, p 5.
49 *AJHR*, 1949 Session I, G-09, p 19.
50 *BPT*, 23 December 1948, p 3.

Chapter 10: The name of the river
1 O. S. Hintz, *Trout at Taupo*, 2nd edn. London: Max Reinhardt, 1964, p 31.
2 'Memorandum from G. F. Yerex to Under-Secretary for Internal Affairs, 27 February 1928', Archives NZ, IA 79/105.
3 Provisionally appointed in 1924, the board received its official instructions and began formal functions in April 1926 (see *NZH*, 24 April 1926, p 13, where a photograph of the first board members appears).
4 'Memo from Surveyor-General to Under-Secretary, Department of Internal Affairs, 1 February 1932', Archives NZ, IA 79/105.
5 *EP*, 5 November 1931, p 8.
6 P. J. Gibbons, 'Andersen, Johannes Carl', *Dictionary of New Zealand Biography*, first published in 1996. Te Ara — the Encyclopedia of New Zealand, https://teara.govt.nz/en/biographies/3a15/andersen-johannes-carl.
7 *EP*, 6 November 1931, p 10.
8 Ibid.
9 W. G. Harding and A. J. Wicks, chief draughtsman, *Map of the Thermal Regions of New Zealand*. Wellington: Head Office, Lands and Survey Department, 1930.
10 *AS*, 9 November 1931, p 6.
11 *EP*, 11 November 1931, p 8.
12 *EP*, 14 November 1931, p 12. A map showing the proposed Tongariro National Park, prepared in April 1887 by the General Survey Office, shows the Waikato River flowing from the slopes of Mt Ruapehu and proceeding north to a point in the vicinity of the Poutu River emanating from Lake Rotoaira: AJHR, 1887, C-18.
13 *EP*, 14 November 1931, p 12, citing Hochstetter's record compiled in 1859.
14 *EP*, 7 December 1931, p 8.
15 *AS*, 15 December 1931, p 3. Te Heu Heu promoted the iwi's

interests at the political level in various ways, and also sought redress in the courts.
16 *EP*, 12 December 1931, p 14.
17 *AS*, 11 February 1932, p 6.
18 *NZH*, 21 January 1932, p 10. Skinner stated that Bidwill was the first European to ascend the Tongariro River.
19 *NZH*, 11 February 1932, p 10.
20 'Submission by Tongariro Anglers Club to Honorary Geographic Board, 6 June 1945', Archives NZ, IA 1, 79/105.
21 'Report of AA (Auckland), 26 November 1946', Archives NZ, IA 1, 79/105.
22 'Memo G. F. Yerex to Under-Secretary of Internal Affairs, 22 June 1945', Archives NZ, IA 1, 79/105.
23 'Letter from Tongariro Anglers Club Inc to Honorary Geographic Board of New Zealand, 6 June 1945', Archives NZ, IA 1, 79/105.
24 'Memorandum of Honorary Geographic Board, Tongariro—Waikato River', Archives NZ, IA 1, 79/105.
25 *Stratford Evening Post*, 2 April 1932, p 6.

Chapter 11: Neilson, Begg and Hickling

1 *NZH*, 30 May 1923, p 8.
2 G. A. Tait (ed.), *Farms and Stations of New Zealand*. Auckland: Cranwell Publishing, 1958, p 391.
3 *Lord Suffer Me to Catch a Fish*, pp 53–54.
4 Ibid.
5 See Hickling, *Freshwater Admiral*, p 80. Begg's claim of 10,000 Tongariro trout is also referred to by Colonel D. W. Beamish in *Trout and Other Fishing in New Zealand*. London: George Allen and Unwin, 1953, p 79.
6 The film, *New Zealand Rainbow*, was later produced by the RKO group and released worldwide.
7 *NZFSG*, February 1933.
8 *NZFSG*, June 1933.
9 As noted in TALTAC records supplied by secretary, Mary Nisbet.
10 *EP*, 7 February 1934, p 15, records Begg as catching an 11¾lb hen fish in perfect condition on the previous day from the Log Pool.
11 *NZFSG*, June 1932.
12 *EP*, 31 March 1932, p 13.
13 His wartime adventures are described in Harold Hickling, *Sailor at Sea*. Wellington: A. H. & A. W. Reed, 1960. Correspondence and papers are held at the Imperial War Museum, London, NRA 28547.
14 Hickling, *Freshwater Admiral*, p 102.
15 Ibid., p 33.
16 Parsons, John, *Parsons' Glory, a bedside book for anglers*. Auckland: Collins, 1976, p 129.
17 G. B. Hobbs writes in *Fisherman's Country* (London: Geoffrey Bles, 1955) of how he used greased-line (floating-line) technique learned on UK salmon rivers to lure Tongariro rainbows to take.
18 Hickling, *Sailor at Sea*, p 163.

Chapter 12: The Upper Waikato and Tongariro Anglers Club

1 'Population of Town Districts', *New Zealand Official Yearbook*, 1907. Statistics New Zealand digital yearbook collection, https://www.stats.govt.nz/indicators-and-snapshots/digitised-collections/yearbook-collection-18932012/.
2 Waitangi Tribunal, *Turangi Township Report 1995*. Wellington: Brookers Ltd, 1995, p 9.
3 *NZH*, 29 September 1930, p 2.
4 *NZH*, 2 November 1932, p 12.
5 'Memorandum from A. Kean, Conservator of Fish and Game, Rotorua to Under-Secretary, Department of Internal Affairs, 29 July 1933', Archives NZ, IA 1933/76/33.
6 'B. U. Barlow, letter of 12 May 1933 to Minister of Internal Affairs', Archives NZ, IA-1-77/27, Pt 1.
7 'File note of J. Bennett, 7 July 1933, recording discussion with F. E. Thornton', Archives NZ, IA-1-77/27, Pt 1.
8 'Internal Affairs Department file note by J. Bennett, 7 July 1933', Archives NZ, IA-1-77/27, Pt 1.
9 'No Angling Sense' by 'An Angler', *Dom*, 30 May 1934, p 11, voiced these and other concerns.
10 *Dom*, 30 May 1934, p 11.
11 *AS*, 4 June 1934, p 4.
12 Ibid, p 5.
13 'Poor Fishing', *AS*, 4 June 1934, p 5.
14 *ODT*, 29 May 1928, p 13.
15 *NZH*, 12 November 1932, p 12.
16 'Minutes of a meeting between UWTAC deputation and Hon. W. E. Parry, 10 July 1936', p 2, Archives NZ, IA-1-77/27, Pt 1.
17 Rachel Barrowman, 'Heenan, Joseph William Allan 1888–1951'. *Dictionary of New Zealand Biography*, updated 16 December 2003. Te Ara — the Encyclopedia of New Zealand, https://teara.govt.nz/en/biographies/.
18 'Memorandum from J. Heenan to Minister of Internal Affairs, 19 May 1936', Archives NZ, Wellington, I A 1, 77/27 Pt 1.
19 'Internal Affairs Department letter to Upper Waikato and Tongariro Anglers Club, 11 June 1934', Archives NZ, IA 1, 77/27, Pt 1.
20 *NZH*, 7 May 1934, p 11.
21 *EP*, 9 November 1937, p 7.
22 'Submission of Tongariro Anglers Club (Inc) to Honorary Geographic Board of New Zealand, 6 June 1945', Archives NZ, Wellington, I A 1, 79/105.
23 'Minutes of inaugural meeting of Lake Taupo and District Anglers Association, Taumarunui, 28 June 1950'.

Chapter 13: The dry fly revolution

1 *EP*, 10 November 1923, p 11.
2 *NZH*, 7 May 1927, p 12.
3 Parsons, *A Taupo Season*, p 44.
4 Hickling, *Freshwater Admiral*, p 218.
5 Rex Chapman-Taylor, *Joe Frost, Jack England and Bill Hamill*, Alexander Turnbull Library, MS-Papers-8319-05.
6 Hickling, *Freshwater Admiral*, p 219.
7 Ibid, p 220.
8 *NZFSG*, 1 December 1939.
9 *Illustrated Sporting and Dramatic News*, 10 December 1937, p 588.
10 *EP*, 7 February 1934, p 15.
11 *NZFSG*, 1 February 1934, p 2.
12 Ibid, p 3.
13 Ibid, p 5.
14 *NZFSG*, 1 May 1934, p 5.
15 *AS*, 28 March 1935, p 22.
16 *The Auckland Weekly News*, 22 May 1935, p 24.
17 Frank Ziegler, 'Dry-fly in wet-fly waters', *Illustrated Sporting and Dramatic News*, 10 December 1937, p 588.
18 *EP*, 6 June 1934, p 7.

Chapter 14: Tongariro camps and lodges in the interwar years

1 Wright, *Illustrated History of New Zealand*, p 315.
2 'Memorandum for Conservator of Fish & Game, Rotorua',

Archives NZ, Wellington, ACIH, Series 16064, MA 31, 13/23a.
3 'Proclamation of 8 October 1926', *NZG*, p 2895.
4 *NZH*, 12 May 1921, p 9.
5 *Newcastle Morning Herald and Miners' Advocate*, 3 April 1926, p 10.
6 *NZH*, 9 April 1927, p 8.
7 Angela Ballara, 'Te Heuheu Tūkino VI, Hoani', *Dictionary of New Zealand Biography*, first published in 1998. Te Ara — the Encyclopedia of New Zealand, https://teara.govt.nz/en/biographies/4t8/te-heuheu-tukino-vi-hoani.
8 *NZH*, 21 March 1925, p 13.
9 *EP*, 12 November 1930, p 12.
10 *NZH*, 27 February 1936, p 16.
11 *NZH*, 7 April 1936, p 14.
12 *NZH*, 24 April 1936, p 18.
13 *NZH*, 27 February 1936, p 16.
14 *AS*, 18 December 1937, p 17.
15 *AS*, 24 February 1944, p 6.
16 Geoffrey C. Branson, 'Trout-fishing in Lake Taupo (New Zealand)', *Walkabout*, 1935, 1(12) (1 October), p 27.
17 http://tongariroriver.org.nz
18 *NZH*, 30 October 1941, p 8.
19 *EP*, 23 November 1928, p 3.
20 *NZFSG*, 1 July 1929, p 8.
21 *NZH*, 19 November 1926, p 14.
22 'Regulations for Trout-fishing, Taupo District', Supplement to *NZG*, 8 October 1926, p 2896.
23 *AS*, 19 November 1926, p 9.
24 'Notes in connection with the question of whether the Hut Camp is within the chain strip', 1928, Archives NZ, Wellington, ACGO 8333, Record IA1, 2112, IA 79/31.
25 *NZH*, 26 September 1930, p 22.
26 *NZFSG*, 1 July 1929, p 8.
27 *NZH*, 14 May 1928, p 11.

Chapter 15: Three great Tongariro pools

1 Hickling, *Freshwater Admiral*, p 126.
2 *NZO*, November 1967, p 12.
3 McDowall, *Essays of a Fishery Scientist*, p 145.
4 Tony Orman, *Trout and Salmon Sport in New Zealand: An Angling Anthology*.Wellington: A. H. & A. W. Reed Ltd, 1980, p 82.
5 Ibid.
6 *NZH*, 23 May 1940, p 8.
7 *NZH*, 4 May 1928, p 12.
8 Hickling, *Freshwater Admiral*, p 126.
9 Supplement to the *The Auckland Weekly News*, 29 May 1924, p 50.
10 *NZH*, 11 April 1914, p 8.
11 *NZH*, 18 May 1926, p 11.
12 *NZH*, 14 April 1927, p 12.
13 *NZH*, 7 May 1927, p 12.
14 *NZH*, 4 May 1928, p 12.
15 *NZH*, 11 May 1928, p 8.
16 *EP*, 16 November 1926, p 4.
17 *NZFSG*, 1 July 1929, p 9.
18 *NZH*, 7 May 1934, p 11.
19 Ibid.
20 *EP*, 4 October 1939, p 8.
21 *NZH*, 4 May 1928, p 12.
22 *NZFSG*, August 1930, p 13.
23 *NZH*, 6 April 1933, p 12.
24 *NZH*, 3 November 1934, p 14.
25 *AS*, 2 November 1936, p 3.
26 *NZH*, 7 March 1938, p 14.
27 *NZH*, 21 April 1938, p 10.
28 *NZH*, 17 April 1941, p 10.
29 Hickling, *Freshwater Admiral*, p 126.
30 *NZH*, 21 April 1927, p 8.

Chapter 16: World War II years and later

1 P. J. Burstall, *Taupo Fisheries Report*, Archives NZ, Wellington, ACGO, 47/8, at p 11, R10764462.
2 *AS*, 20 April 1940, p 8.
3 F. W. Pickard, *Trout Fishing in New Zealand in Wartime*. New York: G. P. Putnam's Sons, 1940.
4 *NZH*, 26 March 1943, p 2.
5 *NZH*, 30 October 1943, p 8.
6 'Department of Internal Affairs, annual report for year ending 31 March 1941', p 12. *AJHR*, 1941 Session I, H-22.
7 Ibid.
8 *NZH*, 1 November 1944, p 6.
9 *The Nelson Evening Mail*, 31 January 1940, p 4.
10 *NZH*, 3 June 1942, p 2.
11 *NZH*, 13 February 1936, p 13.
12 'Department of Internal Affairs, annual report for year ending 31 March 1946', p 8. *AJHR*, 1946 Session I, H-22.
13 *BPT*, 13 June 1949, p 3.
14 'Letter by Senior Fisheries Officer to Controller of Wildlife, 20 November 1952', Archives NZ, Wellington, ACGO, Box 2099 Pt 1, 77/26/6, R14988473.
15 Ibid.
16 'McNamara to Controller of Wildlife, 27 August 1952', Archives NZ, ACGO, Box 2099 Pt 1, 77/26/6, R14988473.
17 'Marine Department press release', Archives NZ, Wellington, ADOE, 1/2/19, R18393848.
18 D. F. Hobbs, 'The Taupo Experimental Winter Season – 1954; Synopsis of Provisional Findings, 8 December 1954', Archives NZ, Wellington, ACGO, Box 95, 47/8, R 10764462.
19 'Record of discussion of Controller of Wildlife and Senior Fisheries Officer, 4 February 1955', Archives NZ, Wellington, ACGO, Box 95, 47/8, R 10764462.
20 D. F. Hobbs, 'Taupo Experimental Season 1955. Review of Recent Investigational Progress', Archives NZ, Wellington, ACGO, Box 95, 47/8, R 10764462.
21 'Memorandum by D. F. Hobbs to G. F. Yerex, 23 November 1955', Archives NZ, Wellington, ACGO, Box 2099, 77/26/6, R14988473.
22 The disputes over research policy are detailed by Galbreath in *Working for Wildlife*, Ch 7.

Chapter 17: The great flood of 1958

1 *NZH*, 28 January 1907, p 8.
2 *NZH*, 23 March 1933, p 10.
3 *NZH*, 23 February 1935, p 14.
4 NIWA Historic Weather Events Catalogue, https://hwe.niwa.co.nz/event/February_1958_New_Zealand_Flooding.
5 Hickling, *Freshwater Admiral*, p 38.
6 'Draft report by P. Burstall, 21 April 1958', Archives NZ, Auckland, AFKC 25230, 276e, R23533509.
7 R. W. Cumberworth, 'Tongariro river – the Hut Pool', *NZO*, November 1967, p 12.
8 'Report on the Tongariro River', *NZO*, April 1958, p 17.

Chapter 18: The Tongariro power development scheme

1 *NZH*, 4 August 1964, p 2.

2 'Notes on fisheries aspects, Tongariro river power development', Archives NZ, Wellington, IA 1, W2633, Box 1834, 47/4/1, R12322775.
3 'Extract from conference between departmental officers and representatives of Tongariro and Lake Taupo Anglers Club [and other clubs] held at Turangi on Tuesday 28 May 1957', Archives NZ, Wellington, IA 1, W2633, Box 1834, 47/4/1, R12322775.
4 *EP*, 15 February 1960, p .14.
5 *EP*, 28 October 1960, p 12.
6 *EP*, 24 March 1964, p 16.
7 *NZH*, 30 July 1964, p 8.
8 *RDP*, 28 September 1963, p.
9 Ibid.
10 *Taupo Times*, 2 July 1964, p 1.
11 *RDP*, 28 September 1963, p 11.
12 *NZH*, 27 September 1963, p 16.
13 *NZH*, 15 July 1964, section 1, p 6.
14 *NZH*, 27 September 1963, p 16.
15 *EP*, 24 June 1964, p 32.
16 Ibid, p 10.
17 Ibid, p 32 (quoting the New Zealand Ombudsman).
18 *EP*, 4 July 1964, p 20.
19 *EP*, 7 July 1964, p 19.
20 J. T. Salmon, *Heritage Destroyed – The Crisis in Scenery Preservation in New Zealand*. Wellington: A. H. & A. W. Reed, 1960.
21 *EP*, 7 July 1964, p 19.
22 'Electric power: generation and supply', *New Zealand Official Yearbook*, 1964, Ch 20A.
23 *NZH*, 13 July 1964, section 1, p 6.
24 *Taupō Times*, 2 July 1964, p 1.
25 *NZH*, 6 August 1964, p 5.
26 *NZH*, 27 July 1964, section 1, p 6.
27 *The Auckland Weekly News*, 19 August 1964, p 37.
28 *EP*, 16 September 1964, p 21.
29 *EP*, 22 September 1964, p 18.
30 *NZH*, 30 October 1964, p 1.
31 The Environment Court made consent orders in [2011] NZEnvC 152, fixing a minimum flow for the Whakapapa river and settling many other aspects of the Tongariro power scheme.
32 'Letter from P. J. Burstall, Conservator of Wildlife, 10 June 1966', Archives NZ, Wellington, W2633, 47/4/1, Pt 4, R3497299.

Chapter 19: The effects of the Tongariro hydro-electricity scheme

1 P. J. Burstall, Assistant Controller of Wildlife; *Fisheries implications — Tongariro power scheme*, Archives NZ, Wellington, ACGO, W2633, 47/4/1, Pt 4, Code: R12322775.
2 'Tongariro will be worthwhile fishing river in five or six years', *Taupō Times*, 19 January 1965.
3 Ibid.
4 C. S. Woods, *Fisheries Aspects of the Tongariro Power Development Project*. Fisheries Technical Report 10. Wellington: Marine Department, 1964, p 2.
5 See clause 6 of the Shand agreement, reproduced at Appendix B.
6 R. T. T. Stephens, *Flow Management in the Tongariro River*. Science and Research Series N. 16. Wellington: Department of Conservation, 1989, p 18.
7 Ibid, p 21.
8 *EP*, 13 July 1971, p 3.
9 *EP*, 9 September 1972, p 3.
10 'Tongariro back on old course', *NZH*, 24 May 1965.
11 John E. Martin (ed.), *People, Politics and Power Stations*. Wellington: Electricity Corporation of New Zealand and Historical Branch, Department of Internal Affairs, 1998, 2nd edition, p 232.
12 *EP*, 8 February 1964, p 21.
13 *NZH*, 16 August 1973, p 2.
14 Peter Gould, *The Complete Taupo Fishing Guide*. Auckland: Collins, 1981, p 226.
15 Trevor Cameron, 'Who's killing the Tongariro?', *Turangi Chronicle*, 11 July–8 August 1987.
16 *AS*, 29 April 1988, p A7.
17 Ibid.
18 Stephens, *Flow Management in the Tongariro River*, p 8.
19 'Statement of evidence of John Gibbs to Environment Court',(2000) at 6.2. Copy supplied to author by J. Gibbs.
20 Ibid, 6.3.
21 'Statement of evidence of Glenn Maclean to Environment Court', 2000, at 14.10. Copy supplied to author by G. Maclean.
22 Ibid, at 4.2.
23 Ibid, at 5.1–5.3.
24 Ibid, at 13.13.
25 Ibid, at 13.29.
26 Ibid, at 13.28.
27 Ibid, at 13.23.
28 'Tongariro Power Development Hearing Committee decision, 30 August 2001', para 8.3.2; https://docs.niwa.co.nz/library/public/TPDHearing2001.pdf.
29 Stephens, *Flow Management in the Tongariro River*, at 2.2.
30 Advocates for the Tongariro River, https://www.tongariroriver.org.nz/issues-overview/lake-level-and-river-flow/.
31 *NZH*, 6 June 2006, p A5.
32 *BPT*, 7 September 2011, p B7.
33 Keith Draper, *Mr Hundred Per Cent: Fred Fletcher's Taupo Tales*. Wellington: A. H. and A. W. Reed, 1969.
34 Taupo Trout Fishing Regulations 1950, r 2(1).
35 *TT*, March 2002, Issue 39, p 27.
36 *TT*, July 2002, Issue 40, p 37.
37 Glenn Maclean, 'Restoring the Tongariro', *TT*, July 2011, Issue 63, p 4.
38 *TT*, March 2002, Issue 39, p 33.
39 Michel Dedual, 'Restoring the early run in Taupo Rivers: Act 2', *TT*, January 2012, Issue 64, p 9.
40 Ibid, p 6.
41 *TT*, October 2015, Issue 67, p 29.
42 Ibid, p 32.
43 Ibid, p 31.
44 Ibid, p 33.
45 Advocates for the Tongariro River, *Annual report 2004*, p 12; https://tongariroriver.org.nz/wp-content/uploads/2020/09/annual_report_04-05.pdf.

Chapter 20: The arrival of nymph fishing

1 John Parsons and Bryn Hammond (eds), 'On the Tongariro', *New Zealand's Treasury of Trout and Salmon: An Angling Anthology*. Auckland: The Halcyon Press, 1989, p 356.
2 'Iron Blue' explained nymphing to Canterbury readers in an article in *The Star*, 7 January 1920, p 8.
3 The Rotorua Trout Fishing Regulations 1971, Amendment No, 2 (1973/18).
4 *NZO*, August 1976, p 16.
5 Ibid, p 17.
6 Taupo Trout Fishing Regulations 1971, Amendment No. 5 (1978/32).
7 The Rotorua and Taupo Trout Fishing Regulations 1980, 4th schedule.

8 *TT*, July 1989, Issue 1, p 5.
9 Ibid, p 4.
10 *TT*, March 2002, Issue 39 , p 8.
11 *Anglers Notice for Taupō District* 2017; https://gazette.govt.nz/notice/id/2017-go2571.
12 'The Pink Nasties', *Flyfisher*, Issue 7, October/November 1984, p 36.
13 *Flyfisher*, Issue 6, August/September 1984, p 6.
14 Anglers Notice for Taupō District 2017, r 1(2); https://gazette.govt.nz/notice/id/2017-go2571.
15 *TT*, November 1996, Issue 23, p 14.
16 *TT*, October 2014, Issue 66, p 22.

Chapter 21: Bridges and floods
1 *The Waikato Times*, 10 September 1891, p 3.
2 *NZT*, 24 June 1911, p 10.
3 'Public Works Statement, Minister of Public Works', *AJHR*, 1913 Session 1, D-01, p 61.
4 'Site plan of Tongariro river bridge (1911)', Archives NZ, R253457320. It seems this plan (PWD 25373) is based on one made in 1909 (PWD 29212, Archives NZ, R25345760).
5 *Taupo Times*, 7 January 1955, p. 1.
6 Archives NZ, Wellington, J46 Cor 1954/993.
7 *TT*, Issue 39, March 2002, p 46.
8 Table based on *TT*, Issue 29, November 1998, p 41; *TT*, Issue 45, March 2004, p 43; and *Taupo Fishery Focus*, issue 25, October 2022.
9 *TT*, Issue 18, March 1995, p 26.
10 *TT*, Issue 29, November 1998, p 49.
11 *TT*, Issue 46, July 2004, p 19.
12 *NZH*, 24 September 2004, p D19; 'Fishing: upper Tongariro full of promise', *NZH*, 23 April 2004, p D19.

Chapter 22: Voices for the river
1 *TT*, November 1998, Issue 49, p 49.
2 Advocates for the Tongariro River, 2002–03 annual report, p 32.
3 Ibid, p 14.
4 Advocates for the Tongariro River, 2010 annual report, p 10.
5 Advocates for the Tongariro River, 2004 annual report, p 11.

Chapter 23: The Tongariro National Trout Centre
1 Tongariro National Trout Centre newsletter, Winter 2010, p 2.
2 'Department of Internal Affairs, annual report for year ending 31 March 1928', p 8; *AJHR*, 1928 Session I, H-22.
3 John Parsons, 'A national tribute to trout', *The Fishing Years*, p 123.
4 John Parsons, *Parsons' Passion: A Troutfisher's Year*. Auckland: Halcyon Press, 1990, p 102.
5 *TT*, March 1999, Issue 30, p 25.
6 Tongariro National Trout Centre financial statements for 2007/08, p 6.
7 Tongariro National Trout Centre newsletter, Summer 2009, p 3.
8 John Parsons, *Pye's Kingdom of Huka Lodge*. Taupō: Acacia Bay Books, 1998.

Chapter 24: Trout by numbers
1 *TT*, November 1994, Issue 17, p 32.
2 Ibid, p 19.
3 *TT*, November 1994, Issue 17, p 18.
4 *TT*, October 2015, Issue 67, p 31.
5 *TT*, December 2002, Issue 41, p 11.
6 Dedual, Michel & Pickford, Michael (2018) The demand for angling licences: A case study from Taupō, New Zealand, Fisheries Research, 204. 10.1016/j.fishres.2018.02.023.
7 *TT*, January 2012, Issue 64, p 22.
8 *TT*, July 2009, Issue 59, p 21.
9 Department of Conservation, Waipa trap reports — 2020 winter season; https://www.doc.govt.nz/parks-and-recreation/places-to-go/central-north-island/places/taupo-trout-fishery/fishery-science/waipa-trap-monthly-reports/2020-winter-season/.

Chapter 25: The future of the fishery
1 *ODT*, 8 December 2017, p .
2 'Letter from P. Burstall to Conservator of Wildlife, 20 December 1957', author's personal collection of archived Tongariro material.
3 *TT*, December 2007, Issue 56, p 49.
4 'Iwi scotches trout farming rumours', *RDP*, 22 June 2015, p A9.
5 'Ngāti Tūwharetoa and Te Kotahitanga o Ngāti Tūwharetoa and the Crown, Deed of Settlement of Historical Claims, 8 July 2017', cl 7.33.2.
6 'Huon Aquaculture convicted of environmental breaches', *TasmanianTimes.com*, 4 May 2020; https://tasmaniantimes.com/2020/05/huon-acquaculture-convicted-environmental-breaches/.
7 See *Dorchester Finance Ltd v Ngahuia Ltd* [2010] NZHC 169.
8 *The Waikato Times*, 9 October 1998, p 3.
9 D. Collins, K. Montgomery and C. Zammit, *Hydrological projections for New Zealand rivers under climate change*. Christchurch: National Institute of Water and Atmospheric Research Ltd, June 2018.
10 Ministry for the Environment, *Our Freshwater 2020*. ME 1490. Wellington: Ministry for the Environment, April 2020.
11 R. I. Woolway, E. Jennings, T. Shatwell et al, 'Lake heatwaves under climate change', *Nature*, 2021, 589, 402–407; https://doi.org/10.1038/s41586-020-03119-1.
12 *TT*, December 2002, Issue 41, p 54.

Appendices
1 *NZPD*, 18 July 1946, p 454.
2 *AS*, 26 December 1925, p 8.
3 *NZG*, 29 July 1948, p 962; *BPT*, 30 April 1934, p 3.
4 Allan and Barbara Cooper, *Pools of the Tongariro*, p 8.
5 Wallace Bain and Barry Greig, *Fishing Guide to the Tongariro River and Lake Taupo's Southern Shore*. Fish and Fowl Series No. 3. Wellington: The Wetland Press, 1983.
6 *TT*, July 2004, Issue 46, p 18.
7 'Minutes of New Zealand Geographic Board, 20 April 2016', p 21; https://www.linz.govt.nz/regulatory/place-names/about-new-zealand-geographic-board.
8 *Frameworks of the New Zealand Geographic Board Ngā Pou Taunaha o Aotearoa*, Version 10, April 2018, p 51.
9 Ibid, p 30.

Index

Published in 2023 by David Bateman Ltd,
Unit 2/5 Workspace Drive, Hobsonville, Auckland 0618, New Zealand
www.batemanbooks.co.nz

ISBN 978-1-77689-004-0

A catalogue record for this book is available from the National Library of New Zealand.

PHOTOGRAPHIC CREDITS
Cover photography: Front cover image: W. H. Horton (Taihape) with an 8½lb trout from the Hut Pool. (W. B. Beattie) *The New Zealand Herald* Glass Plate Collection, Auckland Libraries 1370-626-14. Back cover image: Grace's Pool, Tongariro. (W. B. Beattie) *The New Zealand Herald* Glass Plate Collection, Auckland Libraries 1370-626-01. Author image: Taken at a swing bridge over the Tongariro River, near Major Jones Pool.
All photographs are copyright to their respective title holders, with special thanks to the Alexander Turnbull Library, Wellington, Auckland Libraries Heritage Images Collection, Archives New Zealand, Auckland Museum Collections and the Taupō Museum and Art Gallery.

Book design: Nick Turzynski, redinc. book design, www.redinc.co.nz
Printed in China by Toppan Leefung Printing Ltd